HVAC and Refrigeration Preventive Maintenance

Eric Kleinert

Mc
Graw
Hill
Education

New York Chicago San Francisco
Athens London Madrid
Mexico City Milan New Delhi
Singapore Sydney Toronto

Cataloging-in-Publication Data is on file with the Library of Congress.

McGraw-Hill Education books are available at special quantity discounts to use as premiums and sales promotions, or for use in corporate training programs. To contact a representative please visit the Contact Us page at www.mhprofessional.com.

HVAC and Refrigeration Preventive Maintenance

1 2 3 4 5 6 7 8 9 0   DOC/DOC   1 2 0 9 8 7 6 5 4

ISBN   978-0-07-182565-8
MHID   0-07-182565-7

The pages within this book were printed on acid-free paper.

Sponsoring Editor Judy Bass	**Proofreader** Bhavna Goyal
Editorial Supervisor Donna M. Martone	**Indexer** Robert Swanson
Acquisitions Coordinator Amy Stonebraker	**Production Supervisor** Lynn M. Messina
Project Manager Raghavi Khullar Cenveo® Publisher Services	**Composition** Cenveo Publisher Services
Copy Editor Ragini Pandey	**Art Director, Cover** Jeff Weeks

Contents

Acknowledgments

I would like to convey my utmost appreciation to the following companies for their valuable contributions for this book:

- Dwyer Instruments, Inc.
- Extech Instruments, an FLIR Company
- Fieldpiece Instruments, Inc.
- Klein Tools, Inc.
- Tecumseh Compressor Company

I would also like to thank Judy Bass and the staff of McGraw-Hill for giving me the opportunity to write this book.

I would like to thank Matt Larsen for contributing the artistic icons that were used in the book.

I would like to thank my family for putting up with me over the course of time while writing this book and I hope this book will be helpful to all in the future.

About the Author

Eric Kleinert is a professional with over 40 years of experience in commercial and domestic major appliances; refrigeration; and HVAC sales, service, and installation. He has owned and operated a number of major appliance and air conditioning sales and service corporations. As an instructor for a preeminent technical college, he taught adults aspects of preventative and diagnostic services and techniques, which enhanced their ability to better evaluate products and services necessary to maintain commercial and residential climate-control systems, refrigeration, and major appliances. He is the author of *Troubleshooting and Repairing Major Appliances* (McGraw-Hill, 2012), now in its 3rd edition, and a contributing author of *Handbook of Energy Efficiency and Renewable Energy Technologies* (CRC, 2014).

Introduction

Preventive maintenance practices have a significant effect on air conditioning, heating, ice machines, and refrigeration product's energy use, comfort, and health. A good maintenance program maximizes product performance and life expectancy in the most cost-effective manner.

How many times have you read this statement or heard it from a service company or a manufacturer or something similar?

WARNING! *Failure to perform the required maintenance at the frequency specified in the use and care manual will void the warranty coverage on your new air conditioning and/or heating, ice machine, or refrigeration equipment in the event of a failure.*

To ensure economical, trouble-free operation of the product, the following maintenance is required every X number of months. As owners of this equipment, we fail to either read the use and care manual or perform the necessary preventative maintenance until that day comes when our product fails. Then we have to pay a service company to repair the product and pay a lot more money for the repairs.

What is preventive maintenance and why is it important to have it done to your products? Preventive maintenance is a scheduled activity to inspect, adjust, clean, and test your product's performance and to ensure they run efficiently, so the product will have a long service life.

Homeowners and business owners often operate their products until it breaks down, they are practicing reactive maintenance. Reactive maintenance has low upfront costs which degrades the product's performance and reliability. The end result is the homeowner and/or business owner will have high utility bills and a repair bill to replace faulty components including the preventive maintenance bill. In the long run, the costs are high and it lowers the life expectancy of the product. So be proactive, schedule the preventive maintenance on your equipment.

I can only expose you to the knowledge that's needed for proper preventive maintenance in the air conditioning, heating, and refrigeration industry. This will help you save money on unnecessary repairs. If you are a homeowner or business owner, this book can help you decide if you can perform the required preventive maintenance on your product or if it's the one that's best left for the experts. For the beginning maintenance engineer or HVAC/R service technician, what you learn from this book will give you the keys for a rewarding service career. And for the experienced HVAC/R professional, this book will serve as a good resource and refresher on the proper techniques in performing preventive maintenance on air conditioning, heating, ice machine, and refrigeration equipment.

This book is divided into two sections. Section One—Chapters 1 to 8, covers safety practices, tools needed to perform the maintenance procedures, indoor air quality (IAQ) and air filtration, test, adjust, and balance (TAB) air conditioning equipment and ductwork, basic refrigeration cycle, basic electricity and electronics (high and low voltages), gas and oil fuels. Section Two—Chapters 9 to 17, will provide you with maintenance instructions on specific types of cooling, heating, and refrigeration equipment. The type of equipment will include room air conditioners, residential air conditioning and heating, residential refrigeration (refrigerators, freezers), commercial air conditioning and heating, water-cooling towers, self-contained commercial refrigerators and freezers, commercial ice machines, troubleshooting techniques, and where to get help. Please note that the information contained in this book is only intended to enhance, rather than replace, the manufacturer's recommendation on servicing the equipment.

Also, in this section, each preventive maintenance procedure has been rated by a degree of complexity or simplicity. For the owner of the product, you will have to decide whether you can perform the preventive maintenance procedure or if you must call a professional to complete the preventive maintenance procedures on the product. Following are icon/symbols that are used in this chapter to define who should perform the preventive maintenance on the air conditioning, heating, and refrigeration equipment.

Degree of Difficulty Scale

Expert		Call a certified HVAC service Company
Hard		Service technician
Moderate		Business Owner
Easy		Home Owner

CHAPTER

Safety Practices

S*afety* starts with accident prevention. Injuries are usually caused because learned safety precautions are not practiced. In this chapter some tips are listed to help the homeowner, business owner, maintenance engineer, and the heating, ventilation, and air conditioning/refrigeration (HVAC/R) technician to correctly and safely operate and perform maintenance on air conditioning, heating, and refrigeration equipment.

Any person who cannot use basic tools should *not* attempt to install, maintain, or repair any air conditioning, heating, or refrigeration equipment. Any improper preventative maintenance or repairs will create a risk of personal injury, as well as property damage. Call your manager if preventative maintenance procedures are not fully understood.

Everyone should have access to a first aid kit and should know how to properly use the contents of that kit and its location. It is also recommended that you take a first aid course, such as those offered by the American Red Cross (*www.redcross.org*).

Safety Procedures

Individual, electrical, gas, oil, chemical, air conditioning, heating, and refrigeration operating and safety precautions are generally the same for all air conditioning, heating, and refrigeration equipment. Carefully observe all safety cautions and warnings that are posted on the product being worked on. Understanding and following these safety tips can prevent accidents.

Individual Safety Precautions

Protecting yourself from injuries is vital. Before installing, maintaining, or servicing any air conditioning, heating, or refrigeration equipment do the following:

- Wear gloves as sharp edges on air conditioning, heating, or refrigeration equipment can hurt hands. Also, refrigerants are very cold and could damage the skin.

- Wear safety shoes as accidents are often caused by dropping of heavy items, especially on feet that are not protected.

- Avoid loose clothing that could get caught in the air conditioning, heating, or refrigeration equipment while it is operating.

- Remove all jewelry when working on air conditioning, heating, or refrigeration equipment.

- Tie long hair back.
- Wear safety glasses (ANSI Z87 approved) to protect your eyes from flying debris.
- Use proper tools that are clean and in good working condition when performing preventative maintenance on air conditioning, heating, and refrigeration equipment.
- Have ample light in the work area.
- Be careful when handling access panels or any other components that may have sharp edges.
- Avoid placing hands in any area of the air conditioning, heating, or refrigeration equipment that has not been visually inspected for sharp edges or pointed screws.
- Be sure that the work area is clean and dry from water and oils.
- When working with others, always communicate with each other.
- Always ask for help moving heavy objects.
- When lifting heavy air conditioning, heating, or refrigeration equipment, always use your leg muscles and not your back muscles.
- If you have to work in an attic or in an enclosed space, make sure you use the proper respirator or dust mask.

Electrical Safety Precautions

You should know where and how to turn off the electricity to the air conditioning, heating, and refrigeration equipment (*lock out tag out*). For example, know the location of plugs, fuses, circuit breakers, and cartridge fuses in the home or business; label them. If a specific diagnostic check requires that voltage be applied, reconnect electricity only for the time required for such a check, and disconnect it immediately thereafter. During any such check, be sure that no other conductive parts come into contact with any exposed current-carrying metal parts. When replacing electrical parts or reassembling the air conditioning, heating, or refrigeration equipment, always reinstall the wires back on their proper terminals according to the wiring diagram. Then check to be sure that the wires are not crossing any sharp areas, are not pinched in some way, are not laying on hot or cold surfaces, are not between panels, and are not between moving parts that may cause an electrical problem.

These additional safety tips are also important to remember:

- Always use a separate, grounded electrical circuit for each air conditioning, heating, or refrigeration equipment.
- Never use an extension cord for room air conditioners, heaters, or portable self-contained refrigeration equipment.
- Be sure that the electricity is off before working on any air conditioning, heating, or refrigeration equipment; lock out tag out.
- Never remove the ground wire from a three-prong power cord or any other ground wires from the air conditioning, heating, or refrigeration equipment.
- Never bypass or alter any air conditioning, heating, or refrigeration equipment switch, component, or feature.

- Replace any damaged, pinched, or frayed wiring before repairing the air conditioning, heating, or refrigeration equipment.

- Be sure that all electrical connections within the product are correctly and securely connected.

- Call the service manager if you doubt your abilities. When dealing with electricity, there is no leeway for mistakes.

Gas Safety Precautions

As a technician, I have full respect for electricity and gas fuels. Like electricity, natural and liquefied petroleum gas (LP or LPG) can become dangerous if not handled properly. First, locate the gas shutoff valve by the gas product, and know where the main gas shutoff is located before beginning repairs. The fuel supply (natural or LPG), supply lines, pressure regulator, and LP storage tank will need servicing at least once a year to make certain that there are no leaks. All gas products must be kept clean and free of soot.

The following are some additional safety procedures that a homeowner, business owner, or technician must know before working on gas products:

- Never smoke or light a flame when working on gas products.

- Always follow the manufacturer's recommendations as stated in the use and care manual for gas products.

- Keep all combustible materials, such as gasoline, liquids, paint, paper materials, and rags, away from gas products.

- Do not allow pilot lights to go out. The gas fumes will seep into the room, causing a hazardous condition for anyone who lights a match, turns on a light switch, or uses the telephone.

- When working on gas products always have a fire extinguisher nearby.

- When using gas products, always have the proper ventilation according to the manufacturer's specifications.

- Keep ventilation systems clean of debris and obstructions.

- If you smell gas where the product is located, shut off the gas supply and vent the room.

WARNING *If you cannot turn the gas supply off, you must leave the area, go to a neighbor, and telephone the gas company and the fire department.*

- When the flame turns from a blue color to a yellow color, or if soot forms, you could have a condition known as incomplete combustion.

WARNING *If not corrected immediately, you could have carbon monoxide buildup, which could be potentially fatal. Carbon monoxide is a poisonous nonirritating gas that is odorless, colorless, and has no taste. If you breathe in carbon monoxide, it will mix with your blood and prevent oxygen from entering into the blood supply, causing illness or death.*

- All homes and businesses should have smoke and carbon monoxide detectors installed. Batteries should be replaced annually.

- Insulating materials may be combustible. A gas furnace installed in the attic or other insulated space must be kept free and clear of insulating materials.

- Should the gas supply fail to shut off or if overheating occurs, shut off the gas supply to the product first before shutting off the electrical supply.

- Call the service manager if you doubt your abilities. When dealing with gas, there is no leeway for mistakes.

Oil Safety Precautions

Improper installation, adjustment, maintenance, or service of an oil fired furnace can cause injury, property damage, or death. Always refer to the use and care guide, service literature, and the installation instructions for proper handling of this type of equipment.

The following are some safety procedures that a homeowner, business owner, or technician must know before working on oil products:

- Always follow the manufacturer's recommendations as to where the furnace can be installed within the home or business area.

- Keep the area around the furnace free and clear of combustible materials and liquids, especially newspaper and rags.

- Do not use the furnace area for extra storage of personal goods. The area must be kept clean.

- The furnace must be vented through a good chimney or flue to carry combustion products to the outdoors.

- There must be adequate ventilation around the furnace equipment.

- Make sure that there is adequate combustion and ventilation air to the furnace space per manufacturer's specifications.

- Inspect all ductwork for air leaks and proper installation.

- Never mix gasoline and oil to operate the furnace.

- Never start the furnace burner when oil has accumulated.

- Call the service manager if you doubt your abilities. When dealing with oil, there is no leeway for mistakes.

Chemical Safety Precautions

Chemicals are also dangerous. Knowledge of chemical safety precautions is essential at all times. Consult the material safety data sheets (MSDS) for further information on the chemical being used. The following tips are examples of important practices:

- Remove all hazardous materials from the work area.

- Always store hazardous materials in a safe place and out of the reach of children.

- Before turning on products that use water, run all of the hot water taps in the house or business for approximately 5 minutes. This clears out the hydrogen gas that can build up in the water heater and pipes if they have not been used for more than 2 weeks.

- Call the service manager if you doubt your abilities. When dealing with hazardous materials, there is no leeway for mistakes.

Air Conditioning, Heating, and Refrigeration Safety

- Call the service manager to check out the product if the safety of the product is in doubt.

- Only use replacement parts of the same specifications, size, and capacity as the original part. If you have any questions, contact your local parts dealer, your service manager, or the manufacturer of the product.

- Check water and/or gas connections for possible leaks before reconnecting the electrical power supply. Then completely reassemble the product, remembering to include all access panels.

Operating Safety

After repairing the air conditioning, heating, or refrigeration equipment, do not attempt to operate it unless it has been properly reinstalled according to the use and care manual and to the installation instructions supplied by the manufacturer. If these instructions are not available, do not operate the product. Call the service manager to check out the reinstallation or ask for a copy of the installation instructions from the manufacturer.

Know where the water and/or gas shutoff valves are located for the product, as well as the house's main water or gas shutoff valve; label them. Following these additional safety tips can also prevent injuries:

- Do not allow children to operate air conditioning, heating, or refrigeration equipment.

- Never allow anyone to operate air conditioning, heating, or refrigeration equipment if they are not familiar with its proper operation.

- When discarding old equipment, remove all doors to prevent accidental entrapment and suffocation.

- The refrigerant in refrigerators, freezers, and air conditioning equipment will have to be recovered by an EPA-certified technician before disposal of the product can begin.

- Instruct the customer to use the air conditioning, heating, or refrigeration equipment only for the job that it was designed to do.

Installation Safety Precautions

The first step in ensuring safety with air conditioning, heating, and refrigeration equipment is to be sure that they are installed correctly. Read the installation instructions and the use and care manual that comes with the product. Observe all local codes and ordinances for electrical, plumbing, gas, and oil connections. Ask your local government agency about these codes. Additional safety tips include the following:

- Carefully observe all safety warnings that are contained in the installation instructions and in the use and care manual.

- The work area should be clear of unnecessary materials so that there is plenty of room to work on the air conditioning, heating, and refrigeration equipment.

- The air conditioning, heating, and refrigeration equipment should be protected from the weather, as well as from freezing or overheating.

- The air conditioning, heating, and refrigeration equipment should be correctly connected to its electric, water, gas, oil, drain, and/or exhaust system. It should also be electrically grounded.

Grounding of Air Conditioning, Heating, and Refrigeration Equipment

In 1913, the National Electrical Code (NEC) made grounding at the consumer's home mandatory. Then, in 1962, NEC required all 15-A and 20-A branch circuits have grounding type receptacles. Finally, in 1968, the code required that refrigerators, freezers, and air conditioning equipment be grounded.

The greatest importance of grounding air conditioning, heating, and refrigeration equipment is that it prevents people from receiving shocks from them. However, the major problem associated with the adequate grounding of these products is that many older homes are not equipped with three-prong grounded receptacles. To solve this problem, the consumer must install, or have installed, a properly grounded and polarized three-prong receptacle. A qualified electrician should connect the wiring, and properly ground and polarize the receptacle.

- Remember that safety is the paramount concern, especially when dealing with electricity. Both the technician and the consumer must be aware that it only takes about 100 mA of current to cause death in one second. Here are some additional safety tips:

- Do not install or operate air conditioning, heating, or refrigeration equipment unless it is properly grounded.

- Do not cut off the grounding prong from the room air conditioner or refrigerator or freezer plug.

- Where a two-prong wall receptacle is encountered, it must be replaced with a properly grounded and polarized three-prong receptacle.

- Air conditioning and refrigeration equipment connected to an electrical disconnect box must be properly grounded.

- Call the service manager if you doubt your abilities. When dealing with electricity, there is no leeway for mistakes.

Checking Air Conditioning, Heating, and Refrigeration Equipment Voltage

If it becomes necessary to test air conditioning, heating, or refrigeration equipment with the voltage turned on, observe the following precautions:

- The floor around the air conditioning, heating, or refrigeration equipment must be dry. Water and dampness increase the probability of a shock hazard.
- When using a multimeter, always set the meter correctly for the voltage being checked.
- Handle only the insulated parts of the meter probes.
- Touch components, terminals, or wires with the meter probe tip only.
- Touch the meter probe tips only to the terminals being checked. Touching other components could damage good parts.

Tools Needed for Installation, Repair, and Preventative Maintenance

A basic knowledge of hand tools, electrical tools, and test meters is necessary to effectively complete most installations, repairs, and preventative maintenance. This chapter will cover the basics of each tool. A working knowledge of these tools is a must for the installation, repair, and preventative maintenance of air conditioning, heating, and refrigeration equipment. Always follow safety precautions and manufacturer's recommendations and warnings when handling tools.

Before starting on any type of preventative maintenance or repairs, take the time to put together a toolkit with a selection of good-quality hand tools. A partial list of common hand tools includes:

- *Screwdrivers.* A complete set of flat-blade screwdrivers, ranging from 1/8 to 5/16 in. Handle sizes may vary with the blade dimension. Phillips-tip sizes also vary; the two most common are #1 and #2.

- *Nut drivers.* A complete set is recommended. The common sizes are 3/16, ¼, 5/16, 11/32, 3/8, and ½ in.

- Wrenches
    - *Socket wrenches.* Either 6 or 12 point, ranging in size from 5/32 to 1 in.
    - *Box wrenches.* Common sizes range from ¼ to 1-½ in.
    - *Open-end wrenches.* Common sizes range from ¼ to 15/8 in.
    - *Adjustable wrenches.* The handle size indicates the general capacity. For example, a 4-in. size will take up to a ½-in. nut. A 16-in. handle will take up to a 17/8-in. nut.
    - *Allen wrenches.* Sizes range from 1/16 to about ½ in.

- Claw hammer.

- Adjustable pliers.

- Flashlight.

- Drop-cloth.
- Clamp-on multimeter.

Safety Precautions

Safety starts with accident prevention. Listed in this chapter are some tips to help the homeowner, business owner, or technician when using hand and power tools.

WARNING *Any person who cannot use basic tools should not attempt to install, maintain, or repair any air conditioning, heating, or refrigeration equipment. Any improper installation, preventive maintenance, or repair creates a risk of personal injury and property damage.*

Individual Safety Precautions

Injuries abound when using tools. To be protected from injuries when using hand tools and power tools, do the following:

- Wear gloves.
- Avoid wearing loose clothing when working with power tools.
- Wear safety glasses to protect the eyes from flying debris.
- Use tools according to manufacturer's specifications, and never alter their use.
- Wear safety shoes or boots that are electric shock resistant and static dissipative (SD) rated.
- Wear a respirator or dust mask when working in attics or an enclosed space.

Safety Precautions When Handling Tools

Regardless of which tool is being used, these same rules of care and safety apply:

- Keep tools clean and in good working order.
- Use the tool only for jobs for which it is designed.
- When using power tools, be certain that the power cord is kept away from the working end of the tool.
- If the tool has a shield or guard, be sure it is working properly and remember to use it.
- If an extension cord is used, be sure it is in good working order. Do not use it if bare wires are showing. Also, use a heavy-gauge wire extension cord to ensure adequate voltage for the tool being used.
- Be sure that the extension cord is properly grounded.
- Grip the tool firmly.
- Never use worn-out tools. A worn-out tool has more potential for causing injuries. For example, with a worn-out screwdriver, there is a greater possibility for slips, which could make medical attention necessary.
  - If there is a problem with a power tool, never stick your fingers in the tool. Unplug it first, and then correct the tool's problem.
  - Make sure to use insulated hand tools when working with electricity and electrical components.

FIGURE 2-1   Flat-blade screwdriver.

FIGURE 2-2   Phillips screwdriver.

Screwdrivers

A *screwdriver* is a hand tool used either to attach or remove screws. The two most common types are the flat-blade and Phillips (Figures 2-1 and 2-2). The flat-blade screwdriver is used on screws that have a slot in the screw head. The flat-blade screwdriver is available in many sizes and shapes (Figure 2-3). Always use the largest blade size that fits snugly into the slot on the screw head so that it will not slip off the screw. The screwdriver should never be used as a pry bar or a chisel; it is not designed for that purpose.

A Phillips screwdriver is used to attach or remove screws that have two slots crossing at right angles in the center of the screw head (Figures 2-2 and 2-3).

When using a Phillips screwdriver, exert more pressure downward in order to keep the tool in the slots. Always use the largest Phillips size that fits snugly into the slots, just as with the flat-blade screwdriver.

Never use worn-out screwdrivers when working on air conditioning, heating, or refrigeration equipment. A worn screwdriver may damage the head of the screw. It can also damage the product on which you are working.

Nut Drivers

Many manufacturers use metal screws with a hexagonal head. A *nut driver* is a hand tool similar to a screwdriver, except that the working end of the driver is hexagonal-shaped and fits over a hexagonal nut or a hexagonal bolt head. Each size nut requires a different sized driver (Figure 2-4).

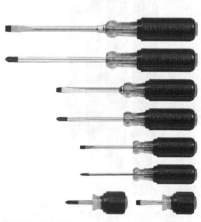

FIGURE 2-3   Combination screwdriver set.

FIGURE 2-4   Hex-nut drivers.

Wrenches

Wrenches are the most frequently used tool. There are many different types and sizes of wrenches. Their purpose is to hold and turn nuts, bolts, cap screws, plugs, and various threaded parts. Wrenches are generally available in five different types (Figure 2-5):

1. Socket wrenches
2. Box wrenches
3. Open-end wrenches
4. Adjustable wrenches
5. Allen or hex wrenches

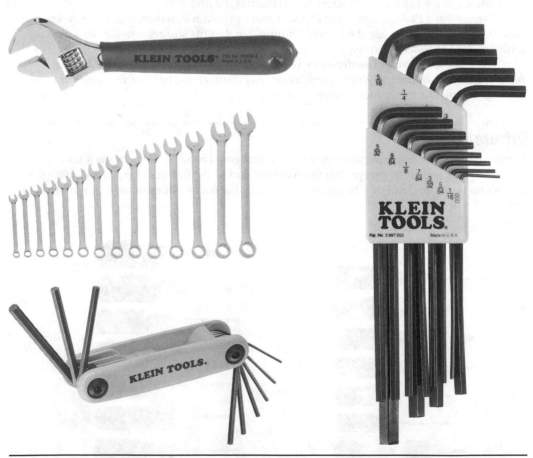

Figure 2-5 Types of wrenches are used to remove and fasten nuts and bolts. They are available in socket, box, open-end, adjustable, and Allen. (*continued*)

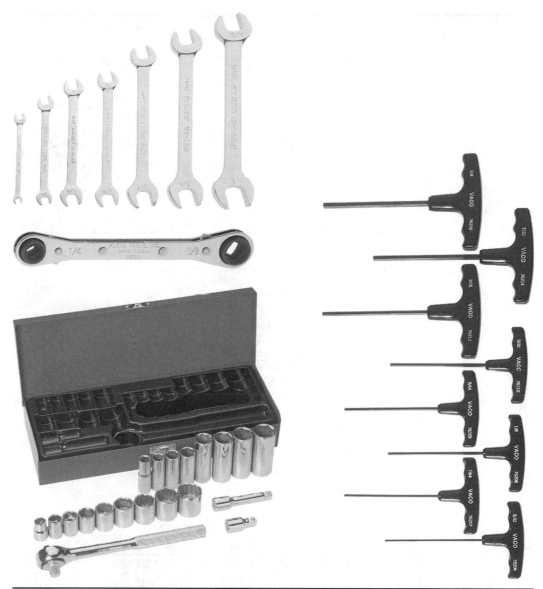

Figure 2-5   Types of wrenches are used to remove and fasten nuts and bolts. They are available in socket, box, open-end, adjustable, and Allen.

Socket wrenches are used to slip over bolt heads, as opposed to other wrenches listed, which are used at right angles to the nut or bolt. This arrangement allows more leverage to be applied to loosen or tighten the nut or bolt (Figure 2-6). Select the size and type of socket to fit the nut with the proper drive size for the load. See Table 2-1 for the proper drive size loading recommendations.

FIGURE 2-6
(*a*) Socket wrench,
(*b*) adjustable wrench,
and (*c*) open-end
wrench.

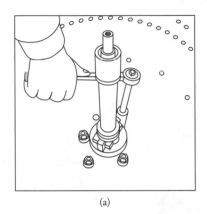

(a)

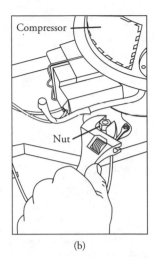

(b)

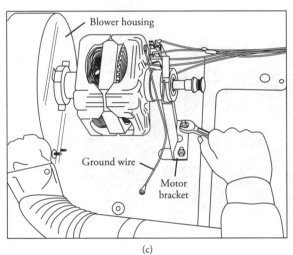

(c)

Hex Size	¼'' Drive	3/8'' Drive	½'' Drive	¾'' Drive	1'' Drive
1/8" to 7/32"	USE	DO NOT USE	DO NOT USE	DO NOT USE	DO NOT USE
¼" to 11/32"	USE	USE	DO NOT USE	DO NOT USE	DO NOT USE
3/8" to 9/16"	USE	USE	USE	DO NOT USE	DO NOT USE
19/32" to 11/16"	DO NOT USE	USE	USE	DO NOT USE	DO NOT USE
¾" to 1"	DO NOT USE	DO NOT USE	USE	USE	DO NOT USE
11/16" to 11/4"	DO NOT USE	DO NOT USE	USE	USE	USE
1 1/2"	DO NOT USE	DO NOT USE	DO NOT USE	USE	USE
1 9/16" to 3 1/2"	DO NOT USE	DO NOT USE	DO NOT USE	DO NOT USE	USE

TABLE 2-1   Drive Size and Hex Size Loading Recommendations

Figure 2-7
While there are many different kinds of hammers, the claw hammer is the one most often used.

Short handle
sledge hammer

Claw
hammer

Ball-peen
hammer

Always choose the correct wrench for the job. *Box wrenches* should be used for heavy-duty jobs and in certain close-quarter situations. *Open-ended wrenches* are useful for medium-duty work or situations where it is impossible to fit a socket or box wrench on a nut, bolt, or fitting from the top. *Adjustable wrenches* help with light-duty jobs, work well with odd-sized nuts and bolts, and are useful where a regular open-end wrench could be used. It is adjustable to fit any size object within its maximum opening. *Allen wrenches* are used for adjusting and removing fan blades or other components that are held in place by Allen set screws. The Allen wrench has a six-pointed flat face on either end.

Hammers

A *hammer* is a hitting tool. There are many sizes and styles of hammers available (Figure 2-7). The most common type used in repairs is the claw hammer. The claw hammer can also be used for prying objects.

Prying Tools

Prying tools are available in many sizes and shapes. The most common are the crowbar, ripping bar, and the claw hammer. The claw hammer is basically used for light-duty work: removing nails and prying small objects. The ripping bar is used for medium-grade work, and the crowbar is used for heavier work.

Pliers

Pliers are one of the most frequently used tools. Depending on the type, they are used for holding or cutting. Generally, they are not made to tighten or unscrew heavy nuts and bolts. They are available in many sizes and shapes (Figure 2-8). Choose a plier that fills a particular need, being careful that it is the proper plier for the job. Some of the most common types of pliers include:

- Slip-joint
- Slip-joint adjustable
- Vise grip
- Needle-nose
- Diagonal cutter

Slip-joint pliers are pliers for everyday tasks (Figure 2-8a). The jaws can be adjusted into two different positions. Do not use them on nuts, bolts, or fittings. They can easily slip and injure both the technician and the device.

Pump pliers, also known as *slip-joint adjustable pliers*, are also used for general jobs (Figure 2-8i). They would be preferred over slip-joint pliers when working on a larger object. The jaws of slip-joint adjustable pliers can be moved into many different positions.

FIGURE 2-8
Pliers are available in many shapes and sizes.

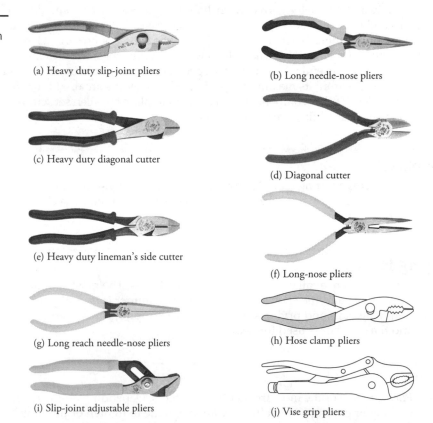

(a) Heavy duty slip-joint pliers

(b) Long needle-nose pliers

(c) Heavy duty diagonal cutter

(d) Diagonal cutter

(e) Heavy duty lineman's side cutter

(f) Long-nose pliers

(g) Long reach needle-nose pliers

(h) Hose clamp pliers

(i) Slip-joint adjustable pliers

(j) Vise grip pliers

The *vise grip plier* is actually four tools in one (Figure 2-8*j*): a clamp, a pipe wrench, a hand vise, and pliers. The lever holds the jaws in one position, allowing the vise to hold up to 1 ton of pressure.

The *needle-nose pliers* are mostly used with electronic, telephone, and electrical work (Figure 2-8*b*, 2-8*f*, and 2-8*g*). Other uses include in confined areas, to form wire loops, and to grip tiny pieces firmly. (The long nose is particularly useful for this latter task.) They are also available with side cutters.

The *diagonal cutting pliers* are used in electrical and electronic work (Figure 2-8*c*, 2-8*d*, and 2-8*e*). They are used for cutting wire and rope.

Cutting Tools

Many different types of tools are used for cutting. The key is to know which tool to use in each situation. Chisels are used for cutting metal and wood. They are made of high-carbon steel, which makes them hard enough to carve through metal (Figure 2-9). These should be used when removing rusted bolts and nuts.

Hacksaws are used for cutting metal (Figure 2-10). The hacksaw consists of a handle, frame, and a blade. The frame is adjustable, so it can accept any length of blade. The blades are available with different numbers of teeth per inch.

A *file* is also a cutting tool. It is used to remove excess material from objects. Files come in a variety of sizes and shapes (Figure 2-11).

Drill bits are also cutting tools. They are designed for cutting holes in metal, wood, and concrete (Figure 2-12).

FIGURE 2-9
Different sizes and shapes of chisels, such as wood chisels, metal chisels, and concrete chisels.

Pin punches   Solid punches

Center punches   Cold chisels

Star drills   Wood chisel

FIGURE 2-10
Hacksaws are used for cutting metal, wood, etc.

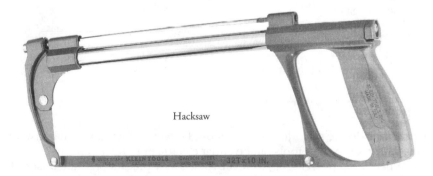

Hacksaw

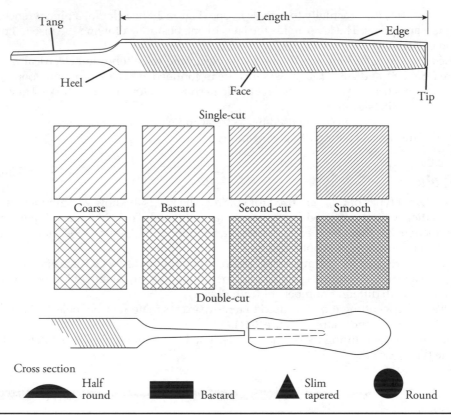

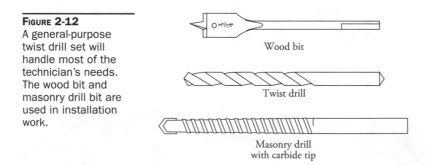

FIGURE 2-11 Files are used to smooth the rough edges of metals, wood, etc.

FIGURE 2-12
A general-purpose twist drill set will handle most of the technician's needs. The wood bit and masonry drill bit are used in installation work.

Wood bit

Twist drill

Masonry drill
with carbide tip

Power Tools

Power tools do the same job as hand tools. However, they do the job faster (Figure 2-13). The most common power tools can be either electric or battery-powered.

When using power tools, there are some safety precautions that must be followed to prevent accidental injury. Always read the use and care manual that comes with each power tool.

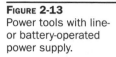

FIGURE 2-13
Power tools with line-
or battery-operated
power supply.

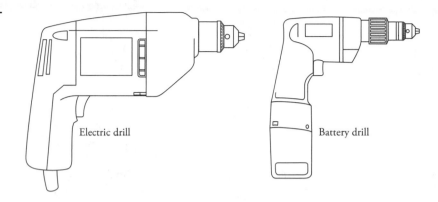

Electric drill   Battery drill

Specialty Tools

These tools are specifically designed for a particular use and are used for in-depth servicing of air conditioning, heating, and refrigeration equipment. For example, special tools are used for the installation and removal of special screws and nuts (Figure 2-14); they are required for security purposes to prevent unauthorized people from tampering with sensitive devices, and they are also used for adjusting switch contacts. Figure 2-15 illustrates some of the many types of specialty tools used in the field. Other specialty tools and their uses will be discussed in later chapters of this book.

Test instruments are important tools used in assisting with the diagnosis of the various problems that arise with air conditioning and refrigeration equipment. Varieties of test meters (Figure 2-16) include the following:

- The analog *volt-ohm-milliammeter* is used for testing the resistance, current, and voltage of the appliance or air conditioner. This type of meter moves a needle along a scale. It is also the most important test meter to have in the tool box.

- The *digital multimeter* displays a readout in numbers, usually on an LCD screen. This type of meter is similar in operation to the analog meter.

- An *ammeter* is a test instrument that is connected into a circuit to measure the current of the circuit without interrupting the electrical current.

- A *wattmeter* is a test instrument used to check the total wattage drawn by an appliance.

- A *temperature tester* is a test instrument used to measure the operating temperatures of the appliance.

Advantages of Digital Meters

- A digital meter offers greater sensitivity, accuracy, and a better resolution in reading measurements.

- Digital meters have a faster readout as compared with analog meters.

Other specialty meters and their uses will be discussed in later chapters of this book.

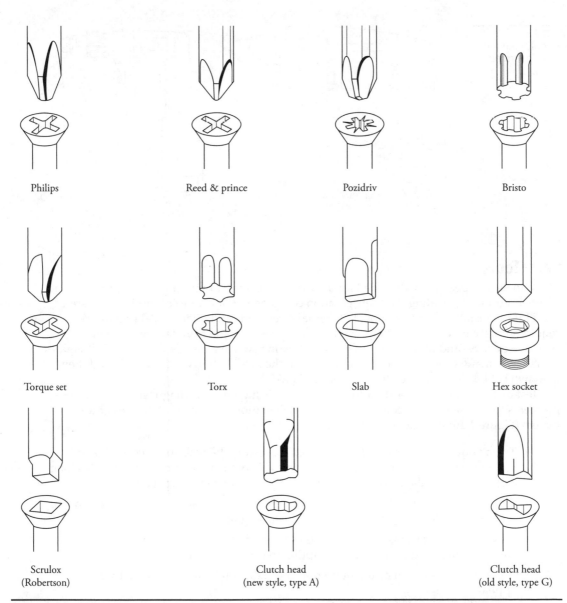

Philips   Reed & prince   Pozidriv   Bristo

Torque set   Torx   Slab   Hex socket

Scrulox   Clutch head   Clutch head
(Robertson)   (new style, type A)   (old style, type G)

FIGURE 2-14   Different types of specialty screws and the types of drivers needed to remove them.

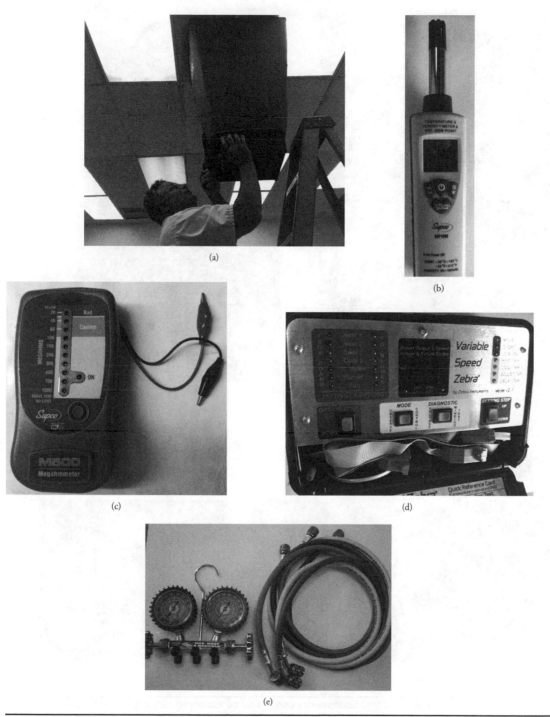

FIGURE 2-15 Special-purpose meters and test instruments used in air conditioning, heating, and refrigeration servicing. (*continued*)

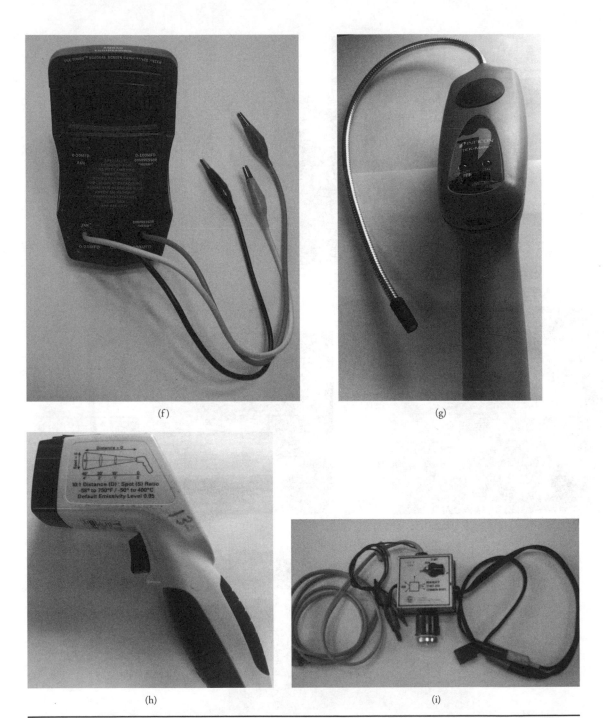

(f)

(g)

(h)

(i)

Figure 2-15 Special-purpose meters and test instruments used in air conditioning, heating, and refrigeration servicing.

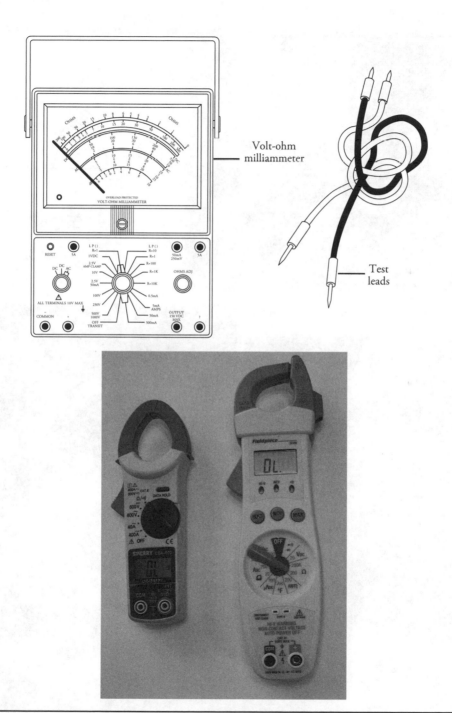

Figure 2-16   Test meters are available in analog or digital readouts, for example, thermometers measure the temperature. (*continued*)

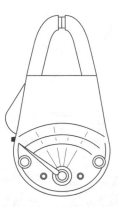

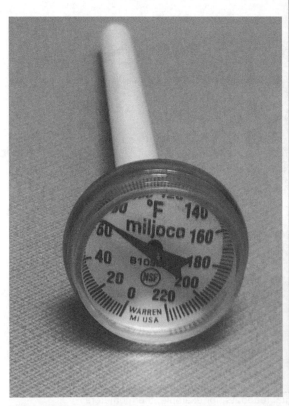

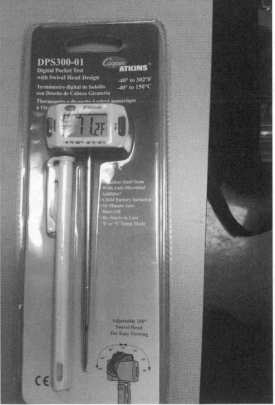

Figure 2-16 Test meters are available in analog or digital readouts, for example, thermometers measure the temperature.

Indoor Air Quality

I s the air in our homes, businesses, and other buildings more polluted than the outside air? Scientific evidence thinks so. People can spend up to 90% of their time indoors. This increases the possibility that many people are at risk to be exposed to air pollution indoors than outdoors. People who are exposed to indoor air pollutants over a longer period of time are more susceptible to the illnesses of air pollution. This includes such groups as the elderly, the young, and the chronically ill who suffers from heart disease and respiratory issues.

This chapter will cover indoor air pollution and how you can prevent it from happening. Being exposed to the risks of environmental pollutants are unavoidable and in most cases, we choose to accept it as a way of life rather than change it. But, we can change the way we breathe the air in our homes and buildings that we occupy most of the time. The buildings that we spend time in have more than one source of air pollutants. The cumulative effects of these other sources of pollutants can be a significant risk by themselves. There are steps most people can take to reduce this source and any future sources of air pollutants from entering buildings.

Sources of Pollutants

In our homes or buildings that we occupy, there are many sources of air pollutants. Some sources are drawn into the building from outdoors, while other pollutants come from within the structure itself. Reducing the exposure to pollutants is a preventable action that can lead to improved outcomes. Some of these air pollutant sources are:

- Asbestos (insulation)
- Building materials
- Cabinetry and furniture (made up pressed wood products)
- Carpet (wet or dry), vinyl, and wood flooring
- Coal
- Fuels and combustion sources
- Furnishings
- Household products (chemicals, including personal products)

- HVAC, humidification, and dehumidification equipment
- Kerosene or gasoline stored in buildings
- Outdoor pollutants including radon and pesticides
- Tobacco products, including second-hand smoke
- Wood

Some of these pollutants can remain in your home or building that you occupy for long periods of time. These pollutants can affect the health of you and your family after exposure or many years later.

Other Sources of Pollutants

Humans breathe about 11,000 to 12,000 in³ of air every day as they go about their daily lives. Unaware of what's in the air, there are over 30 million particles of dust and other pollutants in each cubic foot of air that we breathe. Every day we inhale over 200 million particles of pollutants, many are harmful and in many cases humans are unaware if they will be sick or not. About 50% of all illnesses are related to poor indoor air quality or is aggravated by it. Following are sources of pollutants that are categorized into groups:

- Pollutants outside of the home or building:
  - Automobile and truck exhaust
  - Construction and remolding
  - Dust and dirt airborne
  - Emissions from chemicals and materials stored
  - Industrial pollutants
  - Landfills
  - Lawn, shrub, and tree trimming
  - Pest-control activities including the use of pesticides
  - Pollen, dust, fungal spores
  - Pollutants from roads, parking lots, garages, factories
  - Pollutants in the atmosphere
  - Radon gas
  - Standing water
  - Underground fuel tanks leaking
  - Volatile organic compounds (VOC) from products used in maintenance activities
- Pollutants from HVAC equipment:
  - Dust or dirt that has accumulated on ductwork and other air conditioning components.
  - Microbiological growth in condensate pans, humidifiers, evaporator coils, and ductwork.
  - Improper use of chemical cleaning compounds, sealants, or improper use of biocides.
  - HVAC water towers that are improperly maintained.
- Pollutants from other sources within the office building:
  - Air fresheners and other scented products
  - Elevator equipment rooms and other mechanical systems

- Emissions from bathrooms, kitchen, shops or garages, and cleaning processes
- Emissions from office equipment [VOC, ozone (from copiers)]
- Other supplies and materials

- Pollutants from humans and activities:
  - All cosmetic and perfume odors
  - Burning candles and fireplaces
  - Cleaning chemicals and usage
  - Dust and dirt airborne within building
  - Flaking dead skin, body odors, and hair
  - Personal products such as fragrances and deodorizers
  - Smoking and second-hand smoke
  - Trash odors

- Pollutants from outside maintenance activities:
  - Airborne dirt and dust
  - Emissions from stored chemicals and supplies
  - Improperly maintained cooling towers
  - Pesticides and fertilizers
  - Volatile organic compounds

- Pollutants from buildings and furnishings:
  - Buildings that still contain asbestos.
  - Deteriorated furnishings.
  - Locations within the building that collects or produce dust or fibers, such as carpeting, curtains, or other textiles.
  - Shelving.
  - Standing water.
  - Volatile organic compounds or inorganic compounds that are released from the building or furnishing.
  - Water damage and unsanitary conditions, such as water soaked or water damage furnishings, wet air conditioning ducts, which can contain microbiological growth.

Let's not forget that there are accidental events that can pollute a building also, such as water or other liquid spills, leaks from roofs or piping, microbiological growth due to flooding. When buildings or homes are remodeled or redecorated, there can be pollutants from demolition and rebuilding, including dust and fibers, odors, and VOCs and inorganic compounds from paint, sealants, and adhesives.

Ventilation

Homes and buildings must have adequate outside air ventilation to prevent pollutants from accumulating to levels that can pose health and comfort problems. Buildings that are designed with special mechanical means of ventilation and are constructed to minimize the amount of outside air from entering or leaving the building may have higher pollutant levels than other buildings. Buildings that are considered "leaky" may have pollutant levels higher than normal when some weather conditions greatly reduce outside air from entering the building.

How Outside Air Enters a Home or Building

There are three ways that outside air can enter a home or building you occupy. They are:

1. Infiltration
2. Mechanical ventilation
3. Natural ventilation

Outdoor air will infiltrate a building through openings, joints, cracks in walls, floors, ceilings, and around windows and doors. In natural ventilation, outside air will flow through open windows and doors. Both infiltration and natural ventilation through a building are affected by the temperature differences from the outside air and the inside air and by the velocity of the wind.

Buildings that rely on mechanical ventilation employ such devices as outdoor-vented fans that remove air either from a single room or an entire building, to HVAC equipment and duct work to continuously circulate the indoor air through filters, along with introducing a percentage of outdoor fresh air filtered and circulate both conditioned air sources within the building to strategic points throughout the building. When the exchange rate of the air being replaced is low, the pollutant levels within the building will increase.

There are some solutions to indoor air quality in homes, apartment buildings, and office buildings; you can eliminate or control the sources of pollution. You can increase ventilation and install air cleaning devices. Sometimes by removing a source, altering an activity, unblocking an air supply vent, or opening a window temporarily to increase ventilation can change the air quality.

Your Health and Indoor Air Pollutants

Health effects from exposure from indoor air pollutants may be felt soon after exposure or many years later. Some of these health symptoms from exposure or repeated exposures may include:

- Asthma[1]
- Dizziness
- Fatigue
- Headaches
- Humidifier fever[2]
- Hypersensitivity pneumonitis[3]
- Irritation of the eyes
- Irritation of the nose
- Irritation of the throat

Immediate reactions to indoor air pollutants depend on several factors, age, and preexisting medical conditions. Whether a person reacts to a pollutant depends on the individual sensitivity, and it varies from person to person. Some people can become sensitized to biological pollutants while other people can be sensitized to chemical pollutants.

Certain effects are similar to colds and other viral diseases. It is often difficult to determine if the symptoms were coming from indoor pollutants. The individual must pay attention to the time and place the symptoms occur. If a person leaves the building and the symptoms fade away, and then returns when he goes back into the building, then the problem most likely came from poor indoor air quality from within the building. Then it's time to call an expert to identify the indoor sources that might be causing the symptoms. An investigation will determine if there is inadequate supply of outdoor air, other pollutants, or maybe the effects might be coming from the air conditioning, heating, or humidity conditions prevalent in the building.

Types of Pollutants Found in the Home

In our home we can divide the pollutants into three categories: particles, bioaerosols, and VOCs (Figure 3-1). Particles are in every home and they include:

- Animal dander
- Bacteria and viruses
- Carpet fibers
- Chemical odors
- Dirt
- Dust
- Insulation
- Pet dander
- Pollen
- Smoke

These particles can measure up to about 100 μm (microns) in size and is the diameter of a human hair. Pollen particles are about 8 to 100 μm in diameter; it takes about a dozen small particles to equal the diameter of a human hair. Dust mites are between 30 to 60 μm in diameter. Figure 3-2 illustrates the sizes of these pollutants. Airborne dust particles,

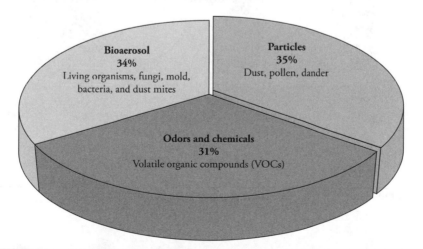

FIGURE 3-1   Categories of pollutants in the home.

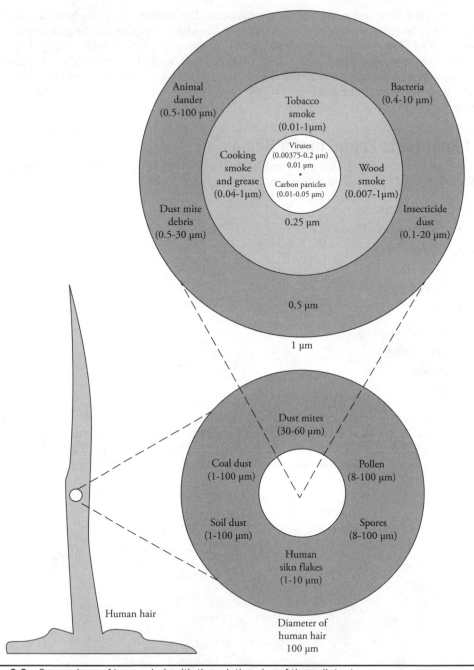

FIGURE 3-2 Comparison of human hair with the relative size of the pollutants.

consisting of fibers and particles float around in our daily environment harmlessly and are no health risk to humans. But, in large concentrations airborne dust particles can cause a risk for health concerns within a home environment.

Bioaerosols are living and non-living microscopic organisms that thrive in your home in warm humid environments. These contaminants travel through the air in our homes and are often invisible. They include:

- Algae
- Animal dander
- Bacteria
- Dust mite allergens
- Fungi
- Mold spores
- Pollen
- Viruses

Bioaerosols are so tiny that they can pass through most air filters and reenter the home living area again. With the right equipment installed, they can be removed and destroyed. To reduce exposure to bioaerosols in several ways:

- Dry off wet surfaces and repair all water leaks within the building.
- In some cases, you would be better off removing the carpeting from the home.
- Install attic ventilation equipment to ventilate the attic and crawl spaces to prevent moisture buildup.
- Keep humidity levels in the home between 30 and 60%.
- Maintain and clean all appliances that come in contact with water. A service technician must inspect and clean air conditioning, heating, humidifying, or dehumidifying equipment. Household appliances including dishwashers, automatic washers, automatic dryers, garbage disposers, and water heaters also need inspecting and cleaning.
- Thoroughly clean and dry carpets.
- Through regular cleaning, you could remove most dust mites, pollens, animal dander, and other allergy-causing agents.
- To prevent biological pollutants in the basements, regularly clean and disinfect any basement floor drains. Use a dehumidifier to keep relative humidity between 30% and 60%.
- Vent exhaust fans from the bathrooms, dryer, and home to the outside of the home.

Volatile organic compounds are vaporous chemicals and odors that have no significant mass that cannot be filtered with conventional methods. Following are some VOCs in the home:

- Building materials and home furnishings
- Cleaning solvents

- Cooking odors

- Fragrances including perfumes, colognes, and hair sprays

- Graphics and craft materials including glues and adhesives, permanent markers

- Household products such as finishes, rug and oven cleaners, paints, lacquers, paint thinners, paint strippers

- Microorganisms that release VOCs, which results in a moldy, musty odor

- Odors from paints and woods

- Pesticides

- Pet odors

Indoor levels of VOCs can be kept to a minimum by selecting products that are low emitting when possible. In addition, VOC concentrations can be diluted through effective ventilation and effective filtration.

Air Filtration

One of the best ways to control pollutants in a building is to have good ventilation and a good air-filtration system installed on the air conditioning or furnace equipment. Air filtration removes unwanted air particles (dust) from the air stream in an air conditioning system within a building. There are two methods of air filtration to purify or filter the air: mechanical filtration and electrostatic filtration. Mechanical filtration consists of air passing through a filter medium that captures the particles from the air passing through the air conditioning equipment and reentering the conditioned space. Types of filter medium used to filter (including fiberglass and pleated) the particles (dust) from the air are rated by a method of testing standards created by American Society of Heating, Refrigerating and Air Conditioning Engineers (ASHRAE).[4] More information on air filters can be obtained from National Air Filtration Association (NAFA).[5]

Electrostatic filtration can be an active or a passive air filtration system. An active electrostatic filter (electronic filter) consists of a high-voltage charge between plates, either a positive or negative charge applied. The dust particles will pass over the plates and will be electrostatically withdrawn from the passing air and captured on the charged plates. This type of filtration system needs constant maintenance performed on it. If maintenance is not performed constantly the efficiency rating will begin to decrease, due to rapid build-up on the collection plate.

Electronic filters emit ozone in the working or living area. Ozone is a pollutant and irritant in high concentrations at ground level. The current guidelines for ground level ozone are high as 51 ppm per hour but levels as high as 51 ppm could provoke health concerns (Table 3-1). People breathing low levels of ozone in the home or office could also become sensitive to ozone. Ozone can trigger a number of health problems including chest pain, shortness of breath, coughing, throat irritations, congestion, and aggravate lung diseases such as asthma, emphysema, and chronic bronchitis.

A passive electrostatic system does not use an applied voltage supply to filter the air particles. This type of filtration system relies on non-conducting dielectric fibers that harbors an electrostatic charge produced from air friction passing through the filter. The air passing through the filter draws out the dust particles electrostatically. The maintenance on

Air Quality Index	Protect Your Health
Good (0–50 ppm)	No health impacts are experienced when the air quality is in this range.
Moderate (51–100 ppm)	Unusually sensitive people should consider limiting prolonged outdoor exertion.
Unhealthy for sensitive groups (101–150 ppm)	The following groups should limit prolonged outdoor exertion: People with lung disease, such as asthma Children and older adults People who are active outdoors
Unhealthy (151–200 ppm)	The following groups should avoid prolonged outdoor exertion: People with lung disease, such as asthma Children and older adults People who are active outdoors Everyone else should limit prolonged outdoor exertion.
Very unhealthy (201-300 ppm)	The following groups should avoid all outdoor exertion: People with lung disease, such as asthma Children and older adults People who are active outdoors Everyone else should limit outdoor exertion.

For more information go to http://www.airnow.gov.

TABLE 3-1   EPA's Guidelines for Air Quality for Ozone

this type of filtration system must be done when the pressure drop (air resistance) exceeds the manufacturer's recommendations and you must clean the filter very well or its efficiency will be decreased after the first use.

As mechanical filters load with particles over time their collection efficiency and pressure drop (air resistance) increase. Over time, the increased pressure drop will inhibit the air flow causing the air conditioning equipment to work harder, losing efficiency, and increasing energy costs. As the pressure drop increases across filter media, it is an indicator that the filters will have to be replaced.

Electrostatic filters will lose their collection efficiency over time when exposed to certain chemicals, aerosols, or high relative humidities. Pressure drop across the filter will increase slowly with an electrostatic filter versus a mechanical filter with similar efficiency. This is a poor indicator that the filters need to be serviced or replaced.

When selecting a filter media for an air conditioner or furnace, keep in mind the differences between the mechanical and electrostatic filters, as it will impact on the filters performance (collection efficiency over time), maintenance requirements (replacement schedules), and total cost.

ASHRAE Standard 52.1 Filter Rating

The American Society of Heating, Refrigeration and Air Conditioning Engineers (ASHRAE) developed Standard 52.1, which contains a number of ratings that have been used for years

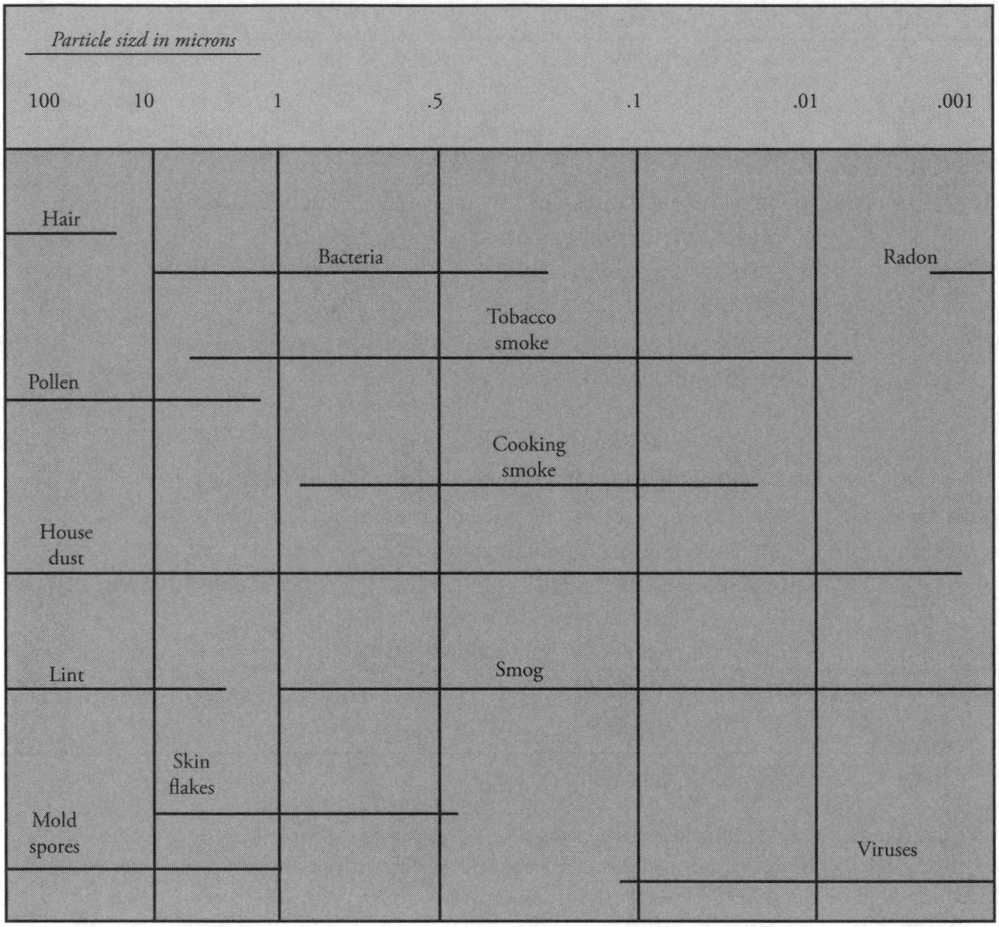

Particle sizd in microns

| 100 | 10 | 1 | .5 | .1 | .01 | .001 |

Hair
Bacteria   Radon
Tobacco smoke
Pollen
Cooking smoke
House dust
Lint   Smog
Skin flakes
Mold spores   Viruses

Figure 3-3   Typical sizes of some common pollutants in a home or office building.

to rate a filter, is based on percentages from 1% to 100%. The most often element mentioned is dust spot efficiency. Standard 52.1 did not mention all particle sizes and disregarded the smaller particles in the 3.0 to 10 µm range (Figure 3-3). These sizes can be very important to the health of the consumer. These smaller particles prompted ASHRAE to come up with Standard 52.2.

Minimum Efficiency Reporting Value

To measure the effectiveness of a filter, ASHRAE Standard 52.2 testing has issued a minimum efficiency reporting value (MERV) rating system. Filters are issued a MERV rating number from 1 to 20, which measures a filter's ability to trap particles ranging from 3.0 to 10 µm (Table 3-2). The higher the MERV rating, the more the filter can filter out the smaller particles efficiently. Residential air conditioning filters have a MERV rating from 1 to 12, commercial and industrial will be higher.

Typical Air-Filter Type	Disposable-Panel Filters, Fiberglass and Synthetic Filters, Permanent Self-Cleaning Filters, Electrostatic Filters, Washable MetalFoam Filters	Pleated Air Filters, Extended-Surface Air Filters, Media-Panel Filters	Non-Supported Bag Filters, Rigid-Box Filters, Rigid-Cell/ Cartridge Filters	Non-Supported Bag Filters, Rigid-Box Filters, Rigid-Cell/ Cartridge Filters MERV 13 Pleats	HEPA Filters, ULPA Filters, SULPA Filters
MERV Std. 52.2	1–4	5–8	9–12	13–16	17–20
Average dust spot efficiency	<20%	<20–35%	40–75%	80–95%+	99.97% 99.99% 99.999%
Average arrestance ASHRAE Std. 52.1	60–80%	80–95%	>95–98%	98–99%	N/A
Particle-size ranges	>10.0 μm	3.0–10.0 μm	1.0–3.0 μm	0.30–1.0 μm	<0.30 μm
Typical air filter applications	Residential, light commercial, equipment protection	Industrial workplace, commercial, paint booths	Industrial workplace, high-end commercial buildings	Smoke removal, general surgery, hospitals and health care	Clean rooms, high-risk surgery, hazardous materials pharmaceutical manufacturing
Types of things filter will trap	Pollen, dust mites, standing dust, spray paint dust, carpet fibers	Mold spores, hair spray, fabric protector, cement dust	Humidifier dust, lead dust, auto emissions, milled flour	Bacteria, most tobaccos, smoke, proplet nuceli (sneeze)	Viruses, radon progeny, carbon dust

TABLE 3-2   ASHRAE Standard 52.2 Filter Rating Chart

ASHRAE Standards 52.1 and 52.2 give consumers an excellent starting point to figure out what type of filter they should use in their air conditioning and heating equipment. The HVAC technician must also be aware of the different types of filters available in the market today. If the wrong type of filter is used, it can decrease the efficiency of the equipment and increase energy usage.

Air Distribution, Air Circulation, and Ductwork

Air distribution is the ability to move air (recirculate the air) in the right quantity, at the right speed, at the designed temperature, controlling the humidity level (moisture level), controlling the noise level of moving air through the ductwork, and cleaning the air from the conditioned space.

The size and location of the ducts, grills, and registers is critical to the proper airflow by the air conditioning equipment. Also, the occupants in the building want the air conditioning equipment to move the air quietly, without a large temperature differential, without drafts from the air moving too fast, and without hot or cold spots from stratified air.

The air-distribution system does not create air; it simply circulates the same air in the building through the building over and over again. The air in a building can be completely circulated through the air conditioning system within 20 minutes. Without proper filtration, without inspecting the ductwork, recirculated air will circulate the same dust particles throughout the building causing indoor air quality (IAQ) problems for the occupants.

Air leaking from the supply ducts can cause the whole house to go into a negative pressure. When a home goes into a negative pressure, the air is being sucked into every crack and crevice of the home, through electrical fixtures, sill plates, windows, and doors. This air is not conditioned air; it is often very hot or cold and with varying degrees of moisture content and dirt. This can cause higher electric bills as well as dust problems. If the air is very humid, the interior of the walls can reach dew point and mold issues can develop where it is not visible, but is potentially dangerous to the occupants.

Air leaking into the return side of the air conditioning system, depending on from where it is being drawn, can have the opposite effect on the home. It will place the structure in a positive pressure. When a home goes positive, it forces the clean, conditioned air out from all the same places, again costing the homeowner more in energy costs. If it is a very dry climate, the moisture level drops in the residence causing low humidity issues such as static electricity, dry itchy skin, nose bleeds, and the like. If in a wet climate, humidity levels can be increased.

Symptoms of Circulation Problems

Some of the symptoms of circulation are:

- Air blows too cold or too hot
- Drafty rooms
- Dust in home
- Home pressure changes when the blower (air handler) turns on
- Humidity high or low
- Moisture around grills
- Musty odors
- Noisy supply or return ducts or grills
- Not cooling or heating in certain rooms
- Not cooling or heating the home
- Sweating ductwork

Solutions of Circulation Problems

Following are some solutions for air-circulation problems:

- Add a UVC light
- Add additional ducts

- Clean ductwork and grills
- Enlarge return grills
- Repair duct leaks
- Replace ducts
- Replace or add insulation
- Variable speed blower (air-handler fan motors)

Endnotes

1. Asthma is a disease that causes the airways of the lungs to swell and narrow, leading to wheezing, shortness of breath, chest tightness, and coughing. In persons who have sensitive airways, asthma symptoms can be triggered by breathing in substances called allergens or triggers. Common asthma triggers include:
   - Animals (pet hair or dander).
   - Certain medicines (aspirin and other NSAIDs).
   - Changes in weather (most often cold weather).
   - Chemicals in the air or in food.
   - Dust mites.
   - Exercise.
   - Mold.
   - Pollen.
   - Respiratory infections, such as the common cold.
   - Strong emotions (stress).
   - Tobacco smoke: Many people with asthma have a personal or family history of allergies, such as hay fever (allergic rhinitis) or eczema. Others have no history of allergies. You can reduce asthma symptoms by avoiding triggers and substances that irritate the airways.
   - Cover bedding with allergy-proof casings to reduce exposure to dust mites.
   - Remove carpets from bedrooms and vacuum regularly.
   - Use only unscented detergents and cleaning materials in the home.
   - Keep humidity levels low and fix leaks to reduce the growth of organisms such as mold.
   - Keep the house clean and keep food in containers and out of bedrooms. This helps reduce the possibility of cockroaches. Body parts and droppings from cockroaches can trigger asthma attacks in some people.
   - If a person is allergic to an animal that cannot be removed from the home, the animal should be kept out of the bedroom. Place filtering material over the heating outlets to trap animal dander. Change the filter in furnaces and air conditioners often.
   - Eliminate tobacco smoke from the home. This is the single most important thing a family can do to help someone with asthma. Smoking outside the house is not enough. Family members and visitors who smoke outside carry smoke residue inside on their clothes and hair. This can trigger asthma symptoms. If you smoke, now is a good time to quit.
   - Avoid air pollution, industrial dust, and irritating fumes as much as possible.

2. Humidifier fever is a disease of uncertain etiology. It shares symptoms with hypersensitivity pneumonitis, but the high attack rate and short-term effects may indicate that toxins, for example, bacterial endotoxins are involved. Onset occurs a few hours after exposure. It is flu like illness marked by fever, headache, chills myalgia, and malaise but without prominent pulmonary symptoms. It normally subsides within 24 hours without residual effects, and a physician is rarely consulted. Humidifier fever has been related to exposure to amoebae, bacteria, and fungi found in humidifier reservoirs, air conditioners, and aquaria.

   Bacterial and fungal organisms can be emitted from impeller and ultrasonic humidifiers. Mesophylic fungi, thermophylic bacteria, and thermophylic actinomycetes all of which are associated with the development of allergic responses have been isolated from humidifiers built into forced air systems as well as separate console units. Airborne concentrations of microorganisms are noted during operation and might be quite high for individuals using ultrasonic or cool mist units.

   Drying and chemical disinfection with bleach or 3% hydrogen peroxide solution are effective remedial measures over a short period, but cannot be considered as reliable maintenance.

   Only rigorous, daily, and end-of-season cleaning regimens, coupled with disinfection, have been shown to be effective. Manual cleaning of contaminated reservoirs can cause exposure to allergens and pathogens.

3. Hypersensitivity pneumonitis is inflammation of the lungs due to breathing in a foreign substance, usually certain types of dust, fungus, or molds. Hypersensitivity pneumonitis may also be caused by fungi or bacteria in humidifiers, heating systems, and air conditioners found in homes and offices.

   Symptoms of acute hypersensitivity pneumonitis may occur 4 to 6 hours after you have left the area where the foreign substance is found, making it difficult to find a connection between your activity and the disease.

   Symptoms may include:
   - Chills
   - Cough
   - Fever
   - Malaise (feeling ill)
   - Shortness of breath
   Symptoms of chronic hypersensitivity pneumonitis may include
   - Breathlessness, especially with activity
   - Cough, often dry
   - Loss of appetite
   - Unintentional weight loss

4. ASHRAE Standard 52.2-1999, 52.2-2007 establishes how filters are rated and tested for efficiency.

5. NAFA Guide to Air Filtration 2001.

Test and Balance

Test, adjust, and balancing (TAB) is a procedure performed to check the HVAC equipment and ductwork to make sure that the system is within design specifications. A system that has not been tested, adjusted, or balanced properly can allow temperature differential within the various rooms of the conditioned space, increased energy operational costs, unhealthy conditions to develop within the workplace or residence, and it could affect the health conditions of the occupants.

When the homeowner or business owner calls a HVAC service company out to inspect, repair, or replace the HVAC equipment, the technician and/or service company must take into account the air duct system before attempting replacement, repairs, or adding refrigerant to a sealed system. If the HVAC system does not have the appropriate air flow from the ductwork or the appropriate return air to the system it cannot be charged correctly after replacement or repairs have been made. The efficiency of the unit decreases and the energy costs to operate the HVAC unit goes up. An HVAC service technician must perform air flow diagnostics on every service call.

In the commercial, industrial, and institutional markets, TAB is more recognized and accepted as a necessary service since the 1960s than it is in the residential market. In the residential market, on existing HVAC equipment, it is also recommended to TAB the equipment and duct system before testing the refrigerant levels. Most residential HVAC equipment is operating in the 55% to 65% range of the equipment rated BTU capacity with increased energy use.[1] After TAB is performed, HVAC equipment will be operating at or above the 90% range of the rated BTU capacity with a reduction in energy use. A home that is properly balanced will help air from stagnating and help reduce indoor air quality (IAQ) issues by conditioning and filtering the indoor air.

Test, adjust, and balance of a HVAC system may have to be done more than once due to seasonal changes or any structural changes in the property. The effectiveness of any HVAC system is completely dependent on the amount of airflow through it. The airflow through a HVAC system is dependent on an adequately sized evaporator (air handler) fan blower; a properly sized, balanced, and designed air duct system; clean evaporator coil(s); and non-restrictive air filters in the air handler and clean condenser coil(s).

TAB Test Instruments

When testing, adjusting, and balancing HVAC equipment, TAB inspectors, HVAC contractors, and HVAC service technicians use specialized test instruments to achieve an outcome of having a well-installed and energy-efficient system that provides conditioned air and comfort to the people using the occupied space. Some of the test instruments used are:

- Analog and digital thermometers
- Psychrometers
- Anemometers
- Analog and digital manometers
- Analog and digital multimeters
- Analog and digital tachometers
- Analog and digital capture hoods

Digital and Analog Thermometers

Digital and analog thermometers (Figure 4-1*a*, 4-1*b*, 4-1*c*, and 4-1*d*) are used for taking temperature measurements to compare return-air and supply-air temperatures in the occupied space and for taking the temperatures of the air entering and leaving at the evaporator coil (air handler) or at the furnace.

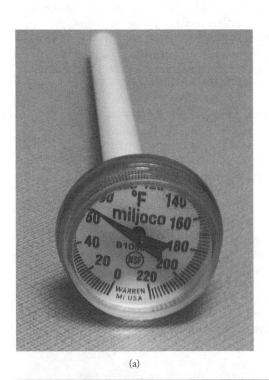

(a)   (b)

Figure 4-1   Different types of temperature measuring instruments: (*a*) analog thermometer; (*b*) digital thermometer; (*c*) IR thermometer; and (*d*) swivel-head clamp meter. (*continued*)

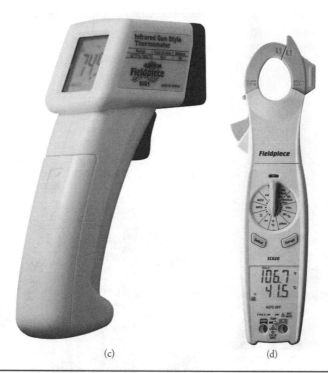

(c)   (d)

Figure 4-1   Different types of temperature measuring instruments: (*a*) analog thermometer; (*b*) digital thermometer; (*c*) IR thermometer; and (*d*) swivel-head clamp meter.

Psychrometers

Psychrometers (Figure 4-2*a*, 4-2*b*, and 4-2*c*) are used to measure wet-bulb temperature, dry-bulb temperature, dew point, and relative humidity. Together these measurements are then plotted on a psychrometric chart to find the enthalpy of air in the measured space or to compare the entering air enthalpy to the leaving air enthalpy across the evaporator coil (in the air handler).[2]

Anemometers

Anemometers are available in several configurations: analog and digital, rotating vane, deflecting vane, and thermal (Figure 4-3). This instrument is primarily used to measure air velocities at registers, grilles, hoods, coils, duct work, and the like. The analog anemometer is a direct reading instrument with a choice of a number of different velocity scales. The digital anemometer has the capability to average the readings and display the velocity readings in feet per minute (fpm).

Analog and Digital Manometers

Manometers (Figure 4-4*a*, 4-4*b*, and 4-4*c*) are used to measure pressure drops, which can be translated into flow rates. This instrument is available in tube types, both U-tube and inclined-vertical. The manometer uses a fluid to represent the difference in pressure

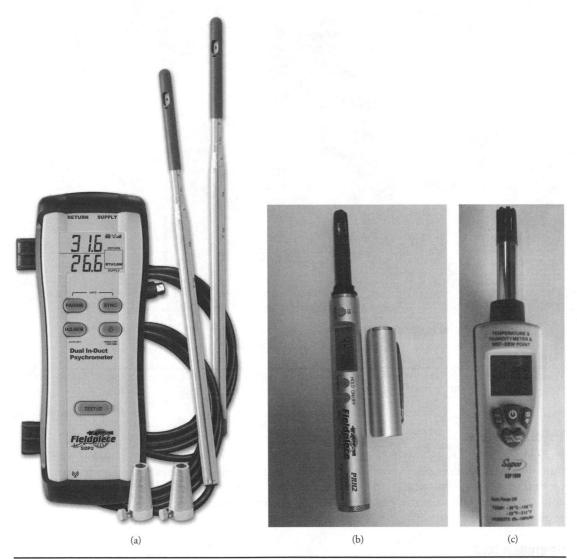

(a)   (b)   (c)

FIGURE 4-2   (a) Dual in-duct psychrometer; (b) pocket psychrometer; and (c) digital hand-held psychrometer.

between two points. Digital manometers can provide very accurate readings at very low pressure differentials, such as across air filters and expansion cooling coils. Digital manometers can automatically adjust for barometric pressure, store readings with recall in average or total numbers, and some models can provide additional functions such as gas pressure, static pressure, and temperature measurements.

Analog and Digital Multimeters

Before the multimeter (Figure 4-5a and 4-5b), a technician would have to have a test meter for each application being tested such as an ohmmeter, voltmeter, wattmeter, capacitor

Figure 4-3

Figure 4-3
In-duct hot-wire
anemometer.

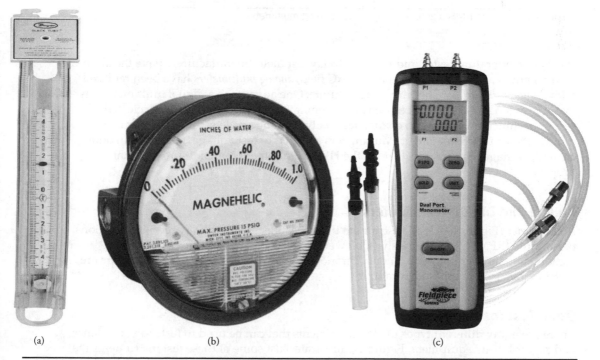

(a)   (b)   (c)

Figure 4-4   (a) U-tube manometer; (b) magnehelic gauge; and (c) digital dual-port differential manometer.

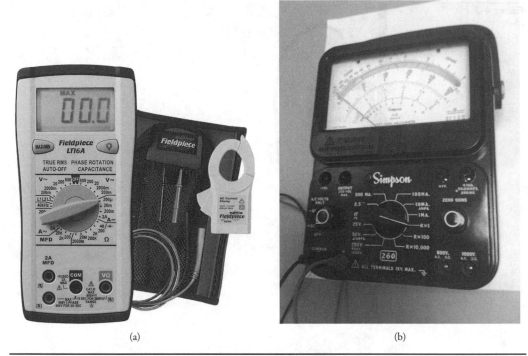

(a)   (b)

Figure 4-5 (a) True RMS digital multimeter; (b) analog multimeter.

checker, temperature tester, and the like. Today test meter manufacturers have the all-in-one meter, known as a multimeter. In the HVAC field, *analog multimeters* have been replaced with *digital multimeters*. They are more accurate. One advantage a digital multimeter has over the analog multimeter, it has an auto ranging feature for each function selected. The service technician will have to change manually the range scale settings on an analog multimeter or risk damaging it. Multimeters are used to test voltage, resistance in ohms, current, wattage, and temperature on the HVAC equipment to make sure it is running properly.

Analog and Digital Tachometers

A *tachometer* (Figure 4-6) is an instrument that measures the rotation speed in revolutions per minute (RPM) of a rotating shaft, blower wheel, or fan blade. The mechanical tachometer, digital photo-tachometer, chronometric tachometer, and the stroboscope are instruments used to measure rotation.

Other Test Instruments

There are many different types of test instruments that can be used to test, adjust, balance, and repair HVAC equipment. Figure 4-7 illustrates just some of these test instruments that are used in the HVAC field. Each test instrument in this chapter and throughout this book

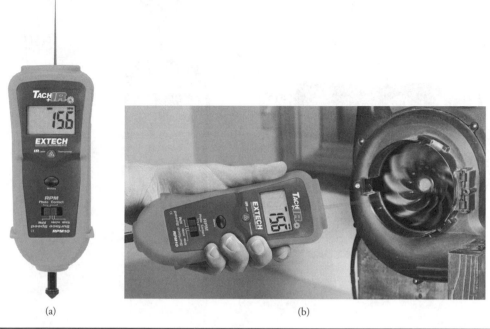

(a)   (b)

Figures 4-6   This digital instrument can be used for (a) contact or (b) non-contact RPM measurement.

(a)   (b)

Figure 4-7   (a) Draft hood; (b) analog manifold refrigerant gauge; (c) digital manifold refrigerant gauge; (d) superheating/subcooling meter; (e) digital megohmmeter; (f) combustion check meter; (g) carbon monoxide detector; (h) infrared refrigerant leak detector; (i) vacuum gauge; (j) variable-speed motor tester; (k) large-vane CFM/CMM anemometer/psychrometer plus CO_2. (continued)

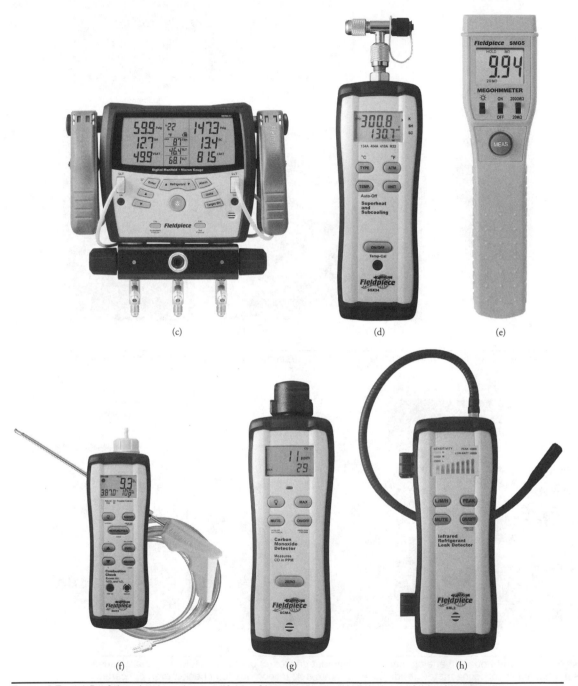

FIGURE 4-7 (a) Draft hood; (b) analog manifold refrigerant gauge; (c) digital manifold refrigerant gauge; (d) superheating/subcooling meter; (e) digital megohmmeter; (f) combustion check meter; (g) carbon monoxide detector; (h) infrared refrigerant leak detector; (i) vacuum gauge; (j) variable-speed motor tester; (k) large-vane CFM/CMM anemometer/psychrometer plus CO_2. (continued)

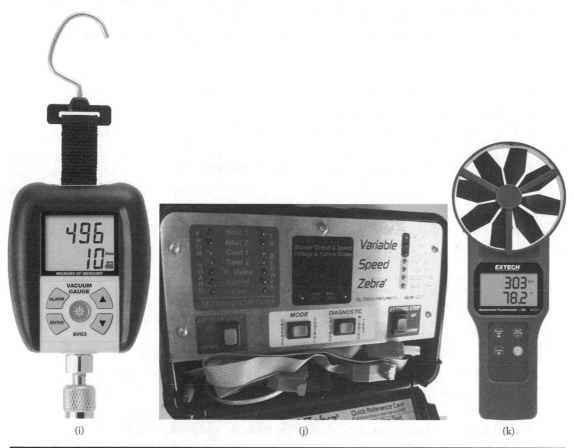

(i)   (j)   (k)

Figure 4-7   (a) Draft hood; (b) analog manifold refrigerant gauge; (c) digital manifold refrigerant gauge; (d) superheating/subcooling meter; (e) digital megohmmeter; (f) combustion check meter; (g) carbon monoxide detector; (h) infrared refrigerant leak detector; (i) vacuum gauge; (j) variable-speed motor tester; (k) large-vane CFM/CMM anemometer/psychrometer plus CO_2.

has a purpose for making sure the HVAC equipment is functioning according to the specifications of the designer's original intent and they will be discussed in later chapters.

Preliminary Procedures for TAB

When preparing a jobsite for TAB, there are some preliminary procedures that must be completed before testing can begin. The preliminary procedures are:

- Prepare all forms and reports for TAB.
- Inspect the jobsite.
- Inspect the air-distribution system.
- Inspect the HVAC equipment.

- Review all drawings and plans for the HVAC equipment.
- Review all electrical characteristics of the HVAC equipment and assure that safety controls are operating.
- Review completely that the HVAC system is ready to be balanced.
- Confirm that the HVAC thermostat has been tested and operational.
- Confirm that HVAC filters are clean.
- Confirm that the building is prepared for testing, for example, windows and doors closed, and the like.
- Confirm that all testing instruments are in good working order.

Preliminary Procedures for the Air Side of HVAC System

Following are the preliminary procedures to the air side of the HVAC system:

- Check duct system, make sure it is intact and properly sealed.
- Set all volume control dampers, variable air boxes to the full open position, unless the HVAC system requires balancing the system by zones.
- Set all fresh air dampers to the minimum position
- Perform and leak test air-duct system.
- Make sure that all access panels and doors are properly secured.
- Verify that ductwork system is installed according to the blueprints and specifications.
- Inspect the ductwork, make sure it is free of debris.
- Check that the fire and smoke dampers are installed and are accessible.
- Check that the terminal boxes, reheat coils, electrical reheat, and the like are installed, functional, and are accessible.
- Check that all other air-distribution devices such as diffusers, grilles, supplies, and the like are installed, functional, and accessible.
- Check and make sure the return air-distribution system has an unobstructed path from each conditioned space back to the HVAC equipment.
- Check and make sure that all air filters and frames are clean, installed correctly, and sealed.
- Check evaporator coils; make sure they are clean and installed correctly.
- Check all drive components, and make sure they are installed correctly.
- Check the blower assembly, and make sure sheaves are properly aligned and tight on the shafts.
- Check and verify the correct rotation of all blowers and fans and check if they are set for the proper speeds.
- Check the V-belts, and make sure they have the correct tension. If the HVAC system has a belt guard, make sure it is properly installed.

- Check and make sure the fan vortex dampers are functional.
- Check and inspect the fan housings, and make sure they are sealed according to the blueprints and specifications.
- Check flexible connections for proper installation.
- Check the fan blower wheel, and make sure it is aligned properly and it has adequate clearance within the housing.
- Check the fan motor, bearings for proper lubrication, if needed
- Check motor voltage, amperage, and make necessary adjustments.

After all preliminary procedures are completed, balancing can commence. When the testing, adjusting, and balancing (TAB) has completed, the final forms and reports will be created and submitted to the customer for review.

Preliminary Procedures for the Water Side of HVAC System

Following are the preliminary procedures to the water side of the HVAC system:

- Check strainers and piping, make sure they are free from debris, clean if necessary.
- Set all balancing devices to full open position.
- On new construction, make sure the construction strainer baskets are replaced with permanent strainer baskets.
- Set mixing valves and the control valves to full coil flow; close coil by-pass valves.
- System is filled to the proper water level and the pressure reducing valve is set.
- Make sure automatic and manual air vents are properly installed and functional.
- Check and make sure that all of the air is purged from the system.
- Check and make sure that water in the expansion tanks are at the proper level.
- Make sure all valves, flow meters, and temperature/pressure tap installed correctly, accessible, and functional.
- Make sure terminal coils are installed, piped correctly, and accessible.
- Make sure pumps are installed and piped correctly, aligned, grounded, and anchored properly.
- Verify pump rotations and proper drive alignments.
- Check vibration isolators properly installed and adjusted.
- Measure and record pump motor voltage and amperage.
- Check pump flow before balancing coils.
- Start air balancing of air system.

After all preliminary procedures are completed, balancing can commence. When TAB has completed, the final forms and reports will be created and submitted to the customer for review.

Preliminary Procedures for a Chiller System

Following are the preliminary procedures for a chiller system:

- Make sure all operating and safety settings for temperature and pressures are correct.
- Make sure the chiller has started and is operating correctly to manufacturer's specifications.
- Take all readings and record them.

After all preliminary procedures are completed, balancing can commence. When TAB has completed, the final forms and reports will be created and submitted to the customer for review.

Visual Inspection of the HVAC Duct System

HVAC ducts can be made out of a variety of materials, including:

- Aluminum
- Copper
- Fiberglass
- Galvanized sheet metal
- Paper fiber
- Plain steel
- Vitrified clay tile

Paper fiber and clay ducts are installed in concrete and aluminum, and copper ducts are typically installed on the outside of a building.

In a forced air duct system, the air comes out of the indoor air handler or furnace in an area called the plenum (Figure 4-8). An extended plenum duct system will have a large rectangular duct (trunk duct or duct transition) connected to this plenum and will extend out in a straight line. The duct that is between the plenum and the air handler or furnace is called the *starting collar.* Ducts that carry air into the room are called supply ducts. Square or round ducts that are connected to and branch off the extended trunk duct (or duct transition) are called *side take-offs.* These supply ducts are connected to register boots or elbows. Changes in direction of the ducts are made by *angle ducts.* A large vertical duct or warm air riser is sometimes called a *stack duct.* A warm supply duct that runs horizontally from the air handler or furnace plenum to a riser is called a *leader.* Dampers may be installed in ducts to control the amount of air moving through the duct. Dampers can be manual or automatically controlled. There are different duct system configurations that are used in commercial and residential installations (Figure 4-9). All of the ducts that carry air back to the air handler or furnace are called *return ducts* (Figure 4-10).

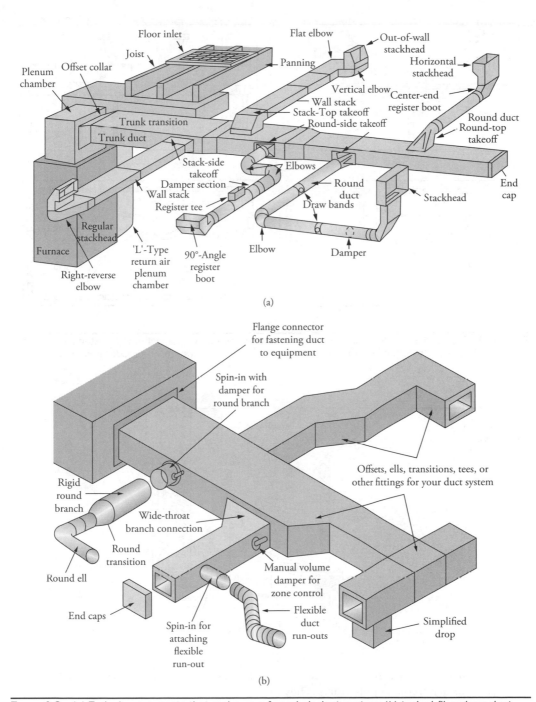

Figure 4-8 (a) Typical components that make up a forced air duct system; (b) typical fiberglass duct system configuration.

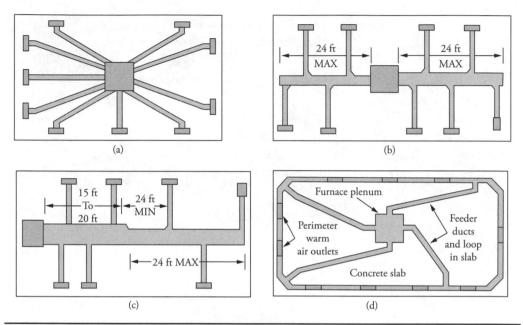

FIGURE 4-9 Typical duct system configurations: (*a*) plenum or radial duct system; (*b*) extended plenum system; (*c*) reducing extended plenum system; (*d*) perimeter-loop system with feeder and loop ducts in concrete slab.

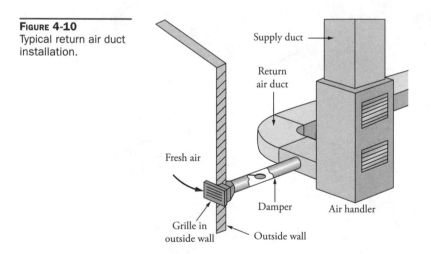

FIGURE 4-10
Typical return air duct installation.

There are three broad types of supply outlet devices (Figure 4-11*a*, 4-11*b*, and 4-11*c*) that are used to attach to the supply ducts. They are:

- Grilles are used to deflect the air up, down, and side to side, depending on the direction in which the louvers are pointed. Grilles are installed on high or low wall locations.

(a)

(b)

(c)

FIGURE 4-11 Typical grilles and registers used to deflect supply air.

- A register is similar to a grille, but a register has a damper to control air flow. They can be installed on walls or floors.
- Diffusers are typically formed in concentric cones or pyramids. They are often installed on ceilings.

FIGURE 4-12
Typical return air grille.

Return air grilles (Figure 4-12) should be located as far away from the supply outlets as possible. Return air grilles are typically located at the bottom of walls near the center of the building.

The service technician should inspect the duct system on every service call to make sure that air is not leaking from or into the duct system. When the technician finds that there is low air flow coming out of one or more supply ducts, the duct system is most likely in need of repair. The following figures illustrate duct work in need of repairs (Figures 4-13 to 4-21). Do you remember the last time that a technician inspected the duct system in your home or business? Maybe it's time for an inspection.

Visual Inspection of the Cooling and Heating system

The manufacturer of the HVAC/R equipment gives a schedule for performing preventive maintenance in the literature supplied with their equipment. These schedules give the

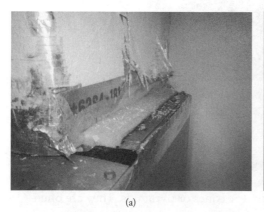

(a)   (b)

FIGURE 4-13   (a) Duct not sealed to the air handler, and (b) air is leaking out of duct system.

FIGURE 4-14
Improper repairs
made to flex duct
system.

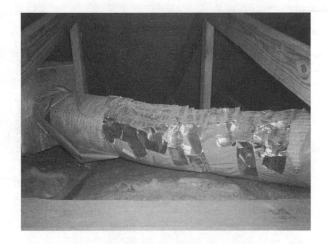

FIGURE 4-15
Return air duct and
air-handler platform in
desperate need of
replacement.

FIGURE 4-16
In this illustration you
can see electrical
wires coming out of
the duct system. This
installation is not
according to the local
building code and
poses a fire hazard.

FIGURE 4-17
Air duct leaking air in attic, causing moisture and mold to collect on the ceiling sheet rock.

FIGURE 4-18
This view was from the inside of the duct. You can see the separation of the other duct.

FIGURE 4-19
Duct separating from the plenum.

FIGURE 4-20
Air leakage at duct seam. Mold is starting to form on duct.

FIGURE 4-21
Leaking air due to torn insulation covering the flex air duct.

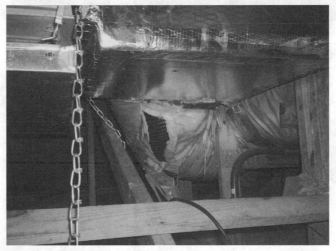

recommended times when to perform the preventive maintenance and also what type of maintenance should be performed on the equipment. In a visual inspection of the HVAC/R equipment, you will look for:

- Evidence of physical damage and leaks
- Dirty filters
- Dirty evaporator and condenser coils
- Signs of overheating
- Unusual sounds or vibrations
- Odors

Look at the following figure and see if you can see any problems (Figure 4-22).

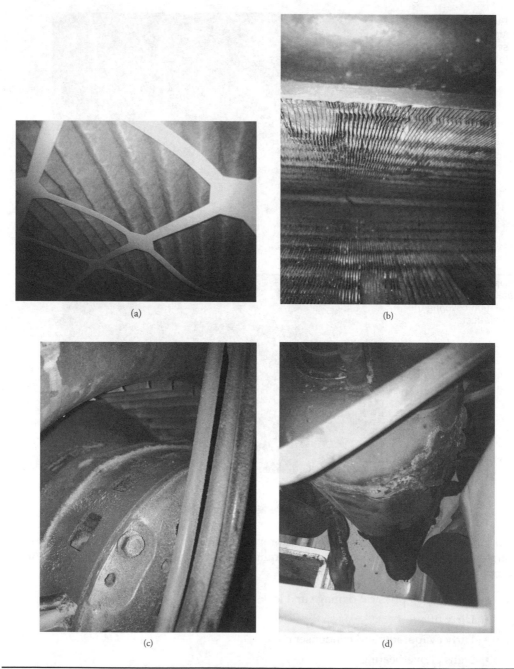

(a)

(b)

(c)

(d)

FIGURE 4-22 (*a*) A dirty filter; (*b*) a dirty evaporator coil; (*c*) dust and dirt accumulating in the evaporator fan motor; (*d*) a compressor rusted and leaking refrigerant and oil; (*e*) condenser coil accumulating sand and dirt in the lower section of the coil; (*f*) liquid line filter drier rusted and leaking refrigerant (high voltage lines are separating from the condenser unit); (*g*) suction line sweating and leaking. (*continued*)

(e)   (f)

(g)

Figure 4-22 (*a*) A dirty filter; (*b*) a dirty evaporator coil; (*c*) dust and dirt accumulating in the evaporator fan motor; (*d*) a compressor rusted and leaking refrigerant and oil; (*e*) condenser coil accumulating sand and dirt in the lower section of the coil; (*f*) liquid line filter drier rusted and leaking refrigerant (high voltage lines are separating from the condenser unit); (*g*) suction line sweating and leaking.

Endnotes

1. BTU—British thermal unit. A BTU is a measure of the amount of cooling or heating that your air conditioning or heating system delivers. A typical ton of cooling or heating is equal to 12,000 BTU, 2 tons = 24,000 BTU, 2 ½ tons = 30,000 BTU, 3 tons = 36,000 BTU, and so on.

2. Psychrometrics is the relationship of the physical and thermal properties of an air vapor mixture.

CHAPTER

Principles of Air Conditioning and Refrigeration

Many excellent books have been written on the subject of air conditioning and refrigeration, and this book will only cover the basics needed to perform preventive maintenance on air conditioning and refrigeration equipment including ice machines and room air conditioners (RACs). This chapter will cover the refrigeration process, but will not cover the replacement procedures of any sealed system components. This is a specialized area, and only an EPA-certified technician can repair the sealed system of a refrigerator or air conditioner (AC). The HVAC/R-certified technician has the proper training, tools, and equipment required to make the necessary repairs to the sealed system. An individual who is not certified to service the sealed system could raise the risk of personal injury, as well as property damage. The uncertified individual could void the manufacturer's warranty on the product if he or she attempts entry into the sealed system.

Introduction to Air Conditioning and Refrigeration

Before the 1900s and mechanical refrigeration, people cooled their foods with ice and snow. The ice and snow were transported long distances and stored in cellars below ground or in icehouses. Refrigeration is the process of removing heat from an enclosed area, thus lowering the temperature in it. Air conditioning and refrigeration began to come to life in the early 1900s. The first practical refrigerating machine was invented in 1834, and it was improved a decade later.

The two most important figures involved with refrigeration and air conditioning were John Gorrie and Willis Carrier. John Gorrie, an early pioneer of refrigeration and air conditioning, was granted a patent in 1851 for his invention of the first commercial refrigeration machine to produce ice. Improvements were made to Gorrie's invention in the late 1880s, and the reciprocating compressor was used commercially in meat and fish plants and for ice production. In 1902, Willis Carrier designed and invented the first modern air conditioning system, and in 1922, Carrier invented the first centrifugal refrigeration machine. Willis Carrier went on to found the company we know today as Carrier Corporation. Toward the end of the 1920s, the first self-contained room air conditioner was introduced.

Refrigerators from the late 1880s until 1929 used ammonia, methyl chloride, or sulfur dioxide as refrigerants. These refrigerants were highly toxic and deadly if a leak occurred. In 1931, the

DuPont Company began producing commercial amounts of "R-12," also known by its trade name *freon*. With safer refrigerants being introduced, the refrigeration and air conditioning markets began to grow. Only half a century later did people realize that these chlorofluorocarbons (CFCs) endangered the earth's ozone layer. Today, refrigerant companies are working harder to produce even safer refrigerants that will not endanger the earth's ozone layer.

Saving the Ozone Layer

High above the earth is a layer of ozone gas that encircles the planet. The purpose of the gas is to block out most of the damaging ultraviolet rays from the sun. Such compounds as chlorofluorocarbons, hydrochlorofluorocarbons (HCFCs), and halons have depleted the ozone layer, allowing more ultraviolet (UV) radiation to penetrate to the earth's surface.

In 1987, the United States, the European Economic Community, and 23 other nations signed the Montreal Protocol on Substances that Deplete the Ozone Layer. The purpose of this agreement was to reduce the use of CFCs throughout the world. To strengthen the original provisions of this protocol, 55 nations signed an agreement in London on June 29, 1990. At this second meeting, they passed amendments that called for a full phase out of CFCs and halons by the year 2000. Also passed at that meeting was the phase out of HCFCs by the year 2020, if feasible, and no later than the year 2040, in any case.

On November 15, 1990, President George H.W. Bush signed the 1990 Amendment to the Clean Air Act, which established the National Recycling and Emissions Reduction Program. This program minimizes the use of CFCs and other substances harmful to the environment, while calling for the capture and recycling of these substances. The provisions of the Clean Air Act are more stringent than those contained in the Montreal Protocol as revised in 1990.

Beginning on July 1, 1992, the Environmental Protection Agency (EPA) developed regulations under Section 608 of the Clean Air Act (the Act) that limit emissions of ozone-depleting compounds. Some of these compounds are known as *chlorofluorocarbons* and *hydrochlorofluorocarbons*. The Act also prohibits releasing refrigerant into the atmosphere while maintaining, servicing, repairing, or disposing of refrigeration and air conditioning equipment. These regulations also require technician certification programs. A sales restriction on refrigerant is also included, whereby only certified technicians will legally be authorized to purchase such refrigerant. In addition, the penalties and fines for violating these regulations can be rather severe.

In 1993, Section 605 of the Clean Air Act establishes the phase out framework. This is an accelerated phase out of class II controlled substances (including R-22). Also, limiting production and consumption of R-22, between 2010 and 2020 to the servicing of equipment manufactured prior to January 1, 2010.

By January 1, 2015, The Montreal Protocol requires the United States to reduce its consumption of HCFCs by 90% below the U.S. baseline.

By January 1, 2020, The Montreal Protocol requires the United States to reduce its consumption of HCFCs by 99.5% below the U.S. baseline.

Matter

In order to understand how refrigeration and air conditioners work, it is necessary to understand several basic laws of *matter*. There are currently five states of matter: *solid, liquid, gas, plasma,* and *Bose-Einstein condensate*.

In our discussion of matter, only three of the five states apply to air conditioning and refrigeration (Figure 5-1):

1. Matter in the form of a solid will retain its shape and volume without a container. The molecules within a solid are compressed and bound together, and under normal conditions will not move at all.

2. Matter in the form of a liquid will take the shape of its container, without losing any volume if not under pressure. Light-density fluids, such as water, will eventually lose some of their volume due to evaporation. The molecules within a liquid are spaced apart from each other and are constantly on the move.

3. Matter in the form of a gas will take the shape and volume of its container and will expand to fill the container. The molecules within a gas are spaced far apart from each other.

Change of State

There are five principles regarding how matter can change from one form to another. When matter is heated, cooled, or an increase or decrease of pressure occurs, matter will be transformed to another state:

1. Liquefication occurs when matter changes from a solid to a liquid. For example, an ice cube will begin to change state when heat is applied.

2. Solidification occurs when matter changes from a liquid to a solid. For example, when the temperature of water reaches its freezing point, it will freeze and become a solid.

3. Vaporization occurs when a liquid matter such as water is heated to its boiling point and transforms into a vapor.

4. Condensation occurs when a vapor transforms into a liquid. For example, when water vapor is boiled off into steam in a closed vessel and then allowed to cool down, it will begin to turn back into a liquid.

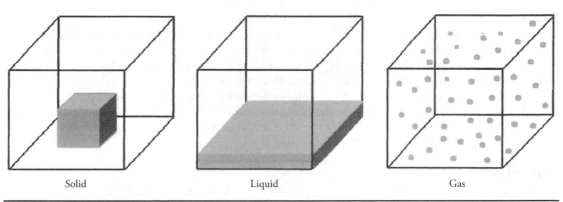

Solid   Liquid   Gas

Figure 5-1   The three states of matter that pertain to air conditioning and refrigeration.

5. Sublimation can occur when a solid matter transforms into a vapor without passing through the liquid state. For example, carbon dioxide in a solid form (dry ice) at atmospheric pressure will transform into a vapor state at normal temperatures, thus bypassing the liquid state.

Law of Thermodynamics

Energy exists in many forms, such as heat, light, mechanical, chemical, and electrical energy. The first law of thermodynamics is the study of that energy. Energy itself is the ability to do work, and heat is only one form of energy—ultimately, all forms of energy end up as heat energy. The first law of thermodynamics states that energy cannot be created or destroyed; it can only be converted from one form to another.

The second law of thermodynamics as it pertains to air conditioning and refrigeration deals with heat energy travel. The law states that heat energy can only travel in one direction—from hot to cold. When two objects are placed together, heat will travel from the warmer object to the cooler object until the temperatures of both objects are equal; only then will heat transfer stop. The rate of travel will depend on the temperature difference between the objects. The greater the temperature difference between the objects, the faster heat will travel to the cooler object.

It's All about Heat

Heat is a form of energy that:

- causes temperatures to rise in an object
- has the capacity to do work
- has the ability to flow from a warmer substance to a cooler one
- can be converted to other forms of energy

Heat energy can be measured with a thermometer. The scales used are Fahrenheit (F) and Celsius (C). Heat intensity has three reference points: freezing point (32°F or 0°C), boiling point (212°F or 100°C), and absolute zero (–470°F or –273°C). This third reference point is believed to be where all molecular action ceases.

Another way to measure heat energy is in units called British thermal units (BTUs). A BTU represents the amount of heat required to raise the temperature of 1 lb of water 1°F at sea level. All refrigeration equipment and air conditioning equipment are rated in terms of how many BTUs of cooling or heating they can remove or put into a given area.

On larger room air conditioners, central air conditioning, and commercial air conditioning it is necessary to calculate the heat load and heat gain of a given area to properly size the equipment. The same is required for large refrigeration equipment and commercial refrigeration equipment.

Methods of Heat Transfer

Heat energy is an important concern in refrigeration and air conditioning. The air conditioning and refrigeration equipment must be able to remove the heat from a given area

FIGURE 5-2
The transfer of heat
from a warmer object
to a cooler object is
known as conduction.

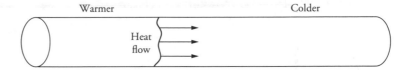

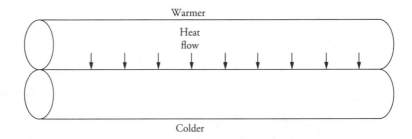

FIGURE 5-2
The transfer of heat
from a warmer object
to a cooler object is
known as conduction.

and transport it to an area outside of the cabinet, room, or building. Heat transfer will take place using one or more of the following basic transfer methods:

- Conduction occurs when two objects are in contact with each other and heat transfers from one object to another (Figure 5-2). The rate of heat transfer will depend on the makeup of the materials used. Conduction will also occur in a single object. For example, if one end of a solid metal rod is heated, the heat will travel toward the cooler end of the rod. Copper and aluminum are good conductors of heat and are used mostly in air conditioners and refrigerators.

- Convection occurs when heat transfers are caused by air movement or fluid movement, whether naturally or by the means of forced air movement (Figure 5-3). Using the refrigerator as an example, the forced air moving over the cold evaporator coil will cause the air to become colder and denser, and it begins to fall to the bottom of the cabinet. In doing so, the air will absorb the heat from the food and air within the refrigerator cabinet. With the heat absorbed in the colder air, it will begin to expand and become lighter; it will rise and, once again, it will be exposed to the colder evaporator coil temperatures, where heat will transfer to the cold coils. This cycle will continue until the temperature requirements are met. Convection also occurs in a liquid when heat transfers in and out of the liquid, as in a refrigerant used in the sealed system of a refrigerator or air conditioner.

- Radiation is the travel of heat energy through the atmosphere by means of radiant waves (light or radio frequency). It travels in a straight path, does not heat the air, and heats only a solid object. For example, if you are standing outside directly in the sunlight, it feels much hotter than if you were to stand in the shade outside. This type of heat transfer comes into play when the refrigerator door is opened or a window covering is opened in an air-conditioned room, allowing heat to enter by radiation. The amount of heat transferred into a refrigerator cabinet or air-conditioned room depends on the temperature difference between the atmosphere and the object.

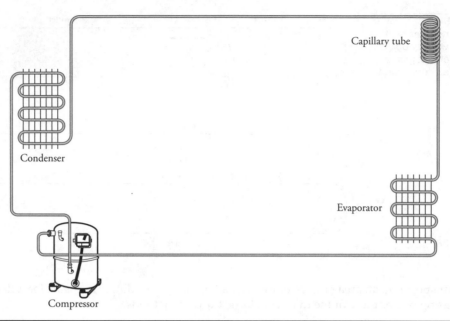

Figure 5-3   The refrigeration cycle.

Refrigerants

Refrigeration involves the process of lowering the temperature of a substance or cooling a designated area. Manufacturers must consider safety, reliability, environmental acceptability, performance, and economics of each refrigerant that is used in refrigeration and air conditioning. The product that absorbs the heat from a substance to be cooled is known as a refrigerant. Refrigerants are classified into two main groups. The first group will absorb the heat by having the refrigerant go through a change of state; the second group involves the absorption of heat without changing the state of the refrigerant. A secondary refrigerant is a refrigerant that will transport the "cold" or "hot" from one location to another (e.g., water). The most widely used method of refrigeration is the vapor compression cycle (refrigeration cycle using a compressor), which relies on the evaporation of a liquid. This type of method is widely used in refrigeration and air conditioning.

Refrigerants are made up of chemical compounds that are used in a sealed refrigeration system of an air conditioner or refrigerator. They absorb the heat by the process of evaporation or boiling the refrigerant, thereby changing it from a liquid state to a gaseous state. Refrigerants have a much lower boiling point than water when it begins to change state. To remove the heat from the refrigerants, they must be able to change back into a liquid state by the process known as condensation.

Refrigerants are classified into three groups according to their flammability:

1. Class 1: non-flammable refrigerant

2. Class 2: moderately flammable refrigerant

3. Class 3: highly flammable refrigerant

A refrigeration system's reliability will also depend on the chemical stability of the refrigerant, its compatibility with the components within the refrigeration circuit, and the type of lubricant used. At all times, the refrigerant should be miscible at various temperatures with the oil used in the compressor to ensure that the oil will circulate throughout the refrigeration circuit and return to the compressor, where it belongs.

R-134a refrigerant is used in domestic refrigerators and domestic freezers, and it is also used in some commercial refrigeration applications. The refrigerant used in air conditioning is R-22, which will be phased out, and R410a will be the replacement refrigerant. There are other substitute refrigerants being used today also.

Remember: You must be certified to service or repair sealed refrigeration systems.

When a sealed system has to be repaired, the acceptable method for recharging the product is the weighed in method. The weight of the refrigerant, usually in ounces (domestic products) and pounds for commercial products, is weighed; this is the most accurate method of refilling the seal system. On larger commercial refrigeration products, superheat and subcooling is the accepted method for recharging the product. Also, these small domestic refrigeration systems are critically charged and are hermetically sealed from the factory, and the manufacturer requires that the products be hermetically resealed again after the repairs are made.

Tables 5-1 to 5-3 illustrate the different types of refrigerants used in modern refrigeration and air conditioning equipment.

Beginning technicians or do-it-yourselfers should check out this Web site before attempting any sealed system repair: *http://www.epa.gov/Ozone/title6/608/general/index.html*.

Refrigerant ASHRAE #	Chemical Name of Refrigerant	Color of Container
R11	Trichlorofluoromethane	Orange
R12	Dichlorodifluromethane	White
R13	Chlorotrifluoromethane	Light blue
R113	Trichlorotrifluoroethane	Dark purple
R114	Dichlorotetrafluoroethane	Navy blue
R12/114	Dichlorodifluromethane, Dichlorotetrafluoroethane	Light gray
R13B1	Bromotrifluoromethane	Pinkish-red
R22	Chlorodifluoromethane	Light green
R23	Trifluoromethane	Light blue gray
R123	Dichlorotrifluoroethane	Light blue gray
R124	Chlorotetrafluoroethane	DOT green
R134A	Tetrafluoroethane	Light blue
R401A	Chlorodifluoromethane, difluoroethane, chlorotetrafluoroethane	Pinkish-red

TABLE 5-1 Refrigerants

Refrigerant ASHRAE #	Chemical Name of Refrigerant	Color of Container
R401B	Chlorodifluoromethane, difluoroethane, chlorotetrafluoroethane	Yellow-brown
R402A	Chlorodifluoromethane, pentafluoroethane, propane	Light brown
R402B	Chlorodifluoromethane, pentafluoroethane, propane	Green brown
R403B	Chlorodifluoromethane, octafluoropropane, propane	Light-gray
R404A	Pentafluoroethane, trifluoroethane, tetrafluoroethane	Orange
R407C	Difluoromethane, pentafluoroethane, tetrafluoroethane	Brown
R408A	Chlorodifluoromethane, trifluoroethane, pentafluoroethane	Medium purple
R409A	Chlorodifluoromethane, chlorotetrafluoroethane, chlorodifluoroethane	Medium brown
R410A	Difloromethane, pentafluoroethane	Rose
R414B	Chlorodifluoromethane, chlorotetrafluoroethane, chlorodifluoroethane, isobutane	Medium blue
R416A	Tetrafluoroethane, chlorotetrafluoroethane, butane	Yellow-green
R417A	Pentafluoroethane, tetrafluoroethane, isobutane	Green
R500	Dichlorodifluoromethane, difluoroethane	Yellow
R502	Chlorodifluoromethane, chloropentafluoroethane	Light purple
R503	Chlorotrifluoromethane, trifluoromethane	Blue-green
R507	Pentafluoroethane, trifluoroethane	Aqua blue
R508B	Trifluoromethane, hexafluoroethane	Dark blue

TABLE 5-1 Refrigerants (*Continued*)

Refrigerant	Chemical Name of Refrigerant
R14	Tetrafluoromethane
R32	Difluoromethane
R41	Fluoromethane
R50	Methane
R125	Pentafluoroethane
R141B	Dichlorofluoroethane
R152A	Difluoroethane
R290	Propane
R600	Butane
R600A	Isobutane

TABLE 5-2 Other Types of Refrigerants

Refrigerant	Chemical Name of Refrigerant
R610	Ethyl ether
R611	Methyl formate
R630	Methyl amine
R631	Ethyl amine
R702	Hydrogen
R704	Helium
R717	Ammonia
R718	Water
R720	Neon
R728	Nitrogen
R732	Oxygen
R740	Argon
R744	Carbon dioxide
R744A	Nitrous oxide
R764	Sulfur dioxide
Refrigerant Categories	
CFC	Chlorofluorocarbons are being phased out due to Montreal protocol agreement
HCH	Hydrocarbons
HCFC	Hydrochlorofluorocarbons—soon to be phased out—production has ceased
HFC	Hydrofluorocarbons are relatively new for vapor compression systems
Azeotropes	A mixture of two or more refrigerants that retains the same composition in vapor form as in liquid form with similar boiling points for a given temperature and pressure and act as one refrigerant
Zeotropes	A mixture of two or more refrigerants that shift in composition during the boiling or condensing process

TABLE 5-2   Other Types of Refrigerants (*continued*)

The Basic Refrigeration Cycle

A service technician needs a good understanding of the refrigeration cycle in order to diagnose air conditioning, refrigeration. Without this understanding, the technician will not be able to diagnose a sealed system problem correctly. Mechanical refrigeration is

Refrigerant ASHRAE #	Trade Name	Refrigerant Components	Components Weight %	Type of Refrigerant	Type of Lubricant
R123		Pure	100%	HCFC	Alkylbenzene or mineral oil
R124		Pure	100%	HCFC	Alkylbenzene or mineral oil
R134A		Pure	100%	HFC	Polyolester oil
R22		Pure	100%	HCFC	Alkylbenzene or mineral oil
R23		Pure	100%	HFC	Polyolester oil
R401A	MP39	R22 / R152A / R124	53% / 13% / 34%	HCFC blended	Alkylbenzene or MO/AB mixture
R401B	MP66	R22 / R152A / R124	61% / 11% / 28%	HCFC blended	Alkylbenzene or MO/AB mixture
R402A	HP80	R125 / R290 / R22	60% / 2% / 38%	HCFC blended	Alkylbenzene or MO/AB mixture
R402B	HP81	R125 / R290 / R22	38% / 2% / 60%	HCFC blended	Alkylbenzene or MO/AB mixture
R403B	ISCEON 69L	R290 / R22 / R218	5% / 56% / 39%	HCFC blended	Alkylbenzene or mineral oil
R404A	HP62, FX70	R125 / R143A / R134A	44% / 52% / 4%	HFC blended	Polyolester oil
R407C	SUVA 9000	R32 / R125 / R134A	23% / 25% / 52%	HFC blended	Polyolester oil
R408A	FX10	R125 / R143A / R22	7% / 46% / 47%	HCFC blended	Alkylbenzene or mineral oil
R409A	FX56	R22 / R124 / R142B	60% / 25% / 15%	HCFC blended	Alkylbenzene or mineral oil
R410A	AZ20, PURON	R32 / R125	50% / 50%	HFC blended	Polyolester oil
R414B	HOT SHOT	R22 / R600A / R124 / R142B	50% / 1.5% / 39% / 9.5%	HCFC blended	Alkylbenzene or mineral oil
R416A	FRIGC FR12	R134A / R124 / R600	59% / 39% / 2%	HCFC blended	Polyolester oil
R417A	NU-22	R125 / R134A / R600A	46.6% / 50% / 3.4%	HFC blended	Alkylbenzene or mineral oil or polyolester oil
R507	AZ50	R125 / R143A	50% / 50%	HFC blended	Polyolester oil
R508B	SUVA 95	R23 / R116	46% / 54%	HFC blended	Polyolester oil

TABLE 5-3 Refrigerant Components, Types, and Lubricants

accomplished by continuously circulating, evaporating, and condensing a fixed supply of refrigerant in a closed loop system. Evaporation occurs at a low temperature and low pressure, while condensation occurs at a high temperature and high pressure. Thus, it is possible to transfer heat from an area of low temperature (i.e., the refrigerator cabinet) to an area of high temperature (i.e., the kitchen area).

There are five components that make up the refrigeration cycle (Figure 5-3 and 5-6a, b, c, and d). They are:

1. compressor
2. condenser coil
3. metering device
4. evaporator coil
5. refrigerant

Let's start the refrigeration cycle at the compressor. The compressor is the heart of the refrigeration cycle, also known as a vapor pump. Once the compressor has started, the suction side of the compressor will draw in a superheated, low pressure, low temperature vapor refrigerant from the evaporator. This vapor is then compressed by the compressor piston into a high pressure, high temperature, and superheated vapor. As the refrigerant exits from the compressor discharge side and enters into the discharge line, it is a high pressure, superheated vapor. The refrigerant will then travel to and enter the condenser coil, giving up some of its heat upon entry.

The beginning portion of the condenser coil is often referred to as the de-superheating section and the refrigerant is in the 100% vapor state. Around the middle of the condenser coil, the refrigerant gives up latent heat and the refrigerant begins to condense. At this point, the refrigerants state is 50% liquid and 50% vapor. As the refrigerant continues to travel and condense, in the condenser coil, the liquid refrigerant increases and the vapor refrigerant decreases. During this time, the refrigerant temperature remains stable. At the bottom of the condenser coil, the refrigerant will become 100% subcooled liquid, and the temperature of the refrigerant is also reduced. The subcooled refrigerant will prevent the refrigerant from boiling. The boiling of the refrigerant at the bottom of the condenser coil is known as flash gas. If this flash gas happens, the capacity of the sealed system would be reduced. To prevent this from happening we must subcool the refrigerant before it reaches the metering device.

The condenser coil's main purpose is to allow the refrigerant to give up its heat. As the refrigerant circulates through the condenser coil, the condenser fan motor will circulate the surrounding air through the condenser coil. As the air passes over the condenser coil, heat will transfer from the refrigerant to the surrounding air.

Inside the condenser coil, the temperature of the high pressure vapor refrigerant will determine the temperature at which condensation begins. Condensation usually begins when the refrigerant is approximately 30°F higher than the surrounding air temperature. At that point, heat will transfer from the refrigerant to the surrounding air. The state of the refrigerant when it enters the liquid line is 100% liquid.

The liquid refrigerant now enters the metering device. The flow of refrigerant into the evaporator is controlled by the pressure differential across the metering device. The metering device for domestic refrigerators, domestic freezers, domestic ice machines, and room air conditioners is called the *capillary tube*. The capillary tube has a very small diameter opening at both ends and is constructed out of copper tubing, with a predetermined length. This type of metering device will control and measure the flow of refrigerant into the evaporator coil. Other types of metering devices that are used in commercial refrigeration, residential and commercial air conditioning, and commercial ice machines are thermostatic expansion valve, automatic expansion valve, electronic expansion valve, float-type metering device, or a fixed bore (orifice or piston) metering device. The location of the metering device is at the end of the liquid line before the inlet of the evaporator coil.

When the refrigerant leaves the metering device, the refrigerant is 75% liquid and 25% vapor. As the high-pressure liquid refrigerant enters the evaporator coil, it is subjected to a much lower pressure due to the suction of the compressor and the pressure drop across the metering device. Thus, the refrigerant tends to expand and evaporate. In order to evaporate, the liquid must absorb heat from the air passing over the evaporator coil. The liquid refrigerant in the evaporator coil will begin to boil and vaporize. This is called latent heat of vaporization. About midway in the evaporator coil the state of the refrigerant is 50% liquid and 50% vapor.

As the refrigerant moves through the evaporator coil it goes from a saturated liquid to a more saturated vapor. When the refrigerant leaves the evaporator coil, it is superheated vapor. This superheated vapor returns to the compressor to begin the cycle over again.

Eventually, the desired air temperature is reached and the thermostat or cold control will break the electrical circuit to the compressor motor and stop the compressor. As the temperature of the air running through the evaporator rises, the thermostat or cold control reestablishes the electrical circuit. The compressor starts, and the refrigeration cycle continues.

It is extremely important to analyze every system completely and understand the intended function of each component before attempting to determine the cause of a malfunction or failure.

Refrigeration Components

As a new technician, you may understand the process of a refrigeration cycle, but you must also understand how each component functions in the refrigeration cycle. Each component has a specific job to assist in maintaining the objective of removing heat from a given area.

Compressor

The compressor (see Figures 5-4*a*, *b*, and *c*) is the heart of the vapor compression system, also known as the pump. All refrigerant circulating within the sealed system must pass through the compressor. This component has two functions within the refrigeration cycle:

- Retrieve the refrigerant vapor from the evaporator, thereby maintaining a constant pressure and temperature.
- Increase the refrigerant vapor pressure and temperature to allow the refrigerant to give up its heat in the condenser coil.

Compressors used for domestic refrigeration and room air conditioners are either reciprocating or rotary in design and are hermetically sealed. Other types of compressors that are used in residential air conditioning, commercial air conditioning, and

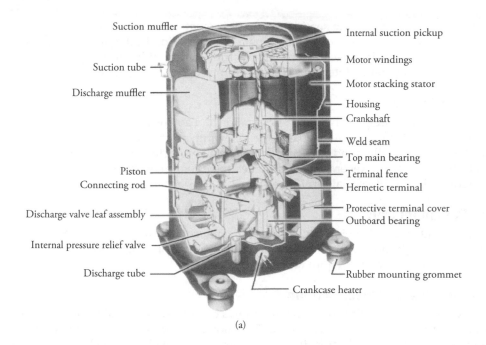

Suction muffler

Suction tube

Discharge muffler

Piston

Connecting rod

Discharge valve leaf assembly

Internal pressure relief valve

Discharge tube

Internal suction pickup

Motor windings

Motor stacking stator

Housing

Crankshaft

Weld seam

Top main bearing

Terminal fence

Hermetic terminal

Protective terminal cover

Outboard bearing

Rubber mounting grommet

Crankcase heater

(a)

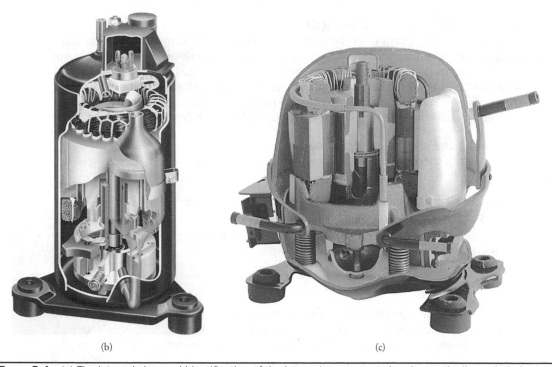

(b)   (c)

FIGURE 5-4 (a) The internal view and identification of the internal components in a hermetically sealed air conditioner compressor; (b) an internal view of a rotary compressor; and (c) an internal view of a compressor.

commercial refrigeration are the semi-hermetic reciprocating, screw, and centrifugal. A hermetically sealed compressor cannot be repaired in the field; it can only be replaced as a complete unit by a certified technician. The electric motor within the compressor can be tested using a multimeter to determine if the windings of the motor are good or bad.

Compressor Safety

In the interest of promoting safety in the refrigeration and air conditioning industry, Tecumseh Products Company has prepared the following information to assist service personnel in safely installing and servicing equipment. This section covers a number of topics related to safety. However, it is not designed to be comprehensive or to replace the training required for professional service personnel.

Trained Personnel Only

Refrigeration and air conditioning devices are extremely complicated by nature. Servicing, repairing, and troubleshooting these products should be done only by those with the necessary knowledge, training, equipment, and certification.

Terminal Venting and Electrocution

Improperly servicing, repairing, or troubleshooting a compressor can lead to electrocution or fire due to terminal venting with ignition. The following precautions can help avoid serious injury or death from electrocution or terminal venting with ignition.

Fire hazard from terminal venting with ignition

Oil and refrigerant can spray out of the compressor if one of the terminal pins is ejected from the hermetic terminal. This *terminal venting* can occur as a result of a grounded fault (also known as a short circuit to ground) in the compressor. The oil and refrigerant spray from terminal venting can be ignited by electricity and produce flames that can lead to serious burns or even death. When spray from terminal venting is ignited, this is called *terminal venting with ignition*. (See Figure 5-5*a*, 5-5*b*, and 5-5*c* for details.)

Terminal venting and electrocution precautions

To reduce the risk of electrocution or serious burns or death from terminal venting with ignition:

- Be alert for sounds of arcing (sizzling, sputtering, or popping) inside the compressor. *Immediately get away* if you hear these sounds.
- Disconnect *all* electrical power before removing the protective terminal cover. Make sure that all power legs are open.

NOTE *The system may have more than one power supply.*

- Never energize the system unless the protective terminal cover is securely fastened and the compressor is properly connected to ground. Figures 5-5*d*, 5-5*e*, and 5-5*f* illustrate the different means of fastening protective terminal covers.
- Never reset a breaker or replace a fuse without first checking for a ground fault (a short circuit to ground). An open fuse or tripped circuit breaker is a strong indication of a ground fault. To check for a ground fault, use the procedure outlined

FIGURE 5-5 (*a*) Compressor with (1) protective terminal cover and (2) bale strap removed to show (3) hermetic terminals; (*b*) close-up view of hermetic terminal showing individual terminal pins with power leads removed; (*c*) close-up view of hermetic terminal after it has vented; (*d*) compressor with (1) protective terminal cover held in place by (2) metal bale strap; (*e*) compressor with (1) protective terminal cover held in place by (2) nut; and (*f*) compressor with (1) snap-in protective terminal cover.

in "Identifying Compressor Electrical Problems" on page 13 in the *Tecumseh Hermetic Compressor Service Handbook* (2011)(located at *www.tecumseh.com*).

- Always disconnect power before servicing, unless it is required for a specific troubleshooting technique. In these situations, use extreme caution to avoid electric shock.

Compressor Motor Starting Relays

A hermetic motor starting relay is an automatic switching device to disconnect the motor start capacitor and/or start winding after the motor has reached running speed. There are two types of motor starting relays used in refrigeration and air conditioning applications: the current responsive type and the potential (voltage) responsive type.

Never select a replacement relay solely by horsepower or other generalized rating. Select the correct relay as specified in the *Tecumseh Electrical Service Parts Guide Book*.

(a)   (b)

Current Type Relay

When power is applied to a compressor motor, the relay solenoid coil attracts the relay armature upward, causing the bridging contact and stationary contact to engage. This energizes the motor start winding.

When the compressor motor attains running speed, the motor main winding current is such that the relay solenoid coil de-energizes, allowing the relay contacts to drop open and disconnecting the motor start winding.

The relay (Figure 5-6a) must be mounted in true vertical position so that armature and bridging contacts will drop free when the relay solenoid is de-energized.

PTC Type Relay

Solid-state technology has made available another type of current sensitive relay: a PTC starting switch (see Figure 5-6b). Certain ceramic materials have the unique property of greatly increasing their resistance as they heat up from current passing through them. A PTC solid-state starting device is placed in series with the start winding and normally has a very low resistance. Upon startup, as current starts to flow to the start winding, the resistance rapidly rises to a very high value, thus reducing the start winding current to a trickle and effectively taking that winding out of operation.

Usage is generally limited to domestic refrigeration and freezers. Because it takes 3 to 10 minutes to cool down between operating cycles, it is not feasible for short-cycling commercial applications.

Compressor Thermal Protectors (Overloads)

All hermetic compressors have either an internal or an external protector (overload) to protect the motor from overheating. The compressor overload in Figure 5-7a is attached to the compressor in Figure 5-7c and 5-7d. The overload is designed to react quickly by means of a bimetal disk to an increase in temperature or an increase in current.

The bimetal disk (Figure 5-7b) consists of two dissimilar metals combined together. Any change in temperature or current draw will cause it to deflect, actuating the switch contacts. When the bimetal cools, the reverse action takes place. The external protector is a

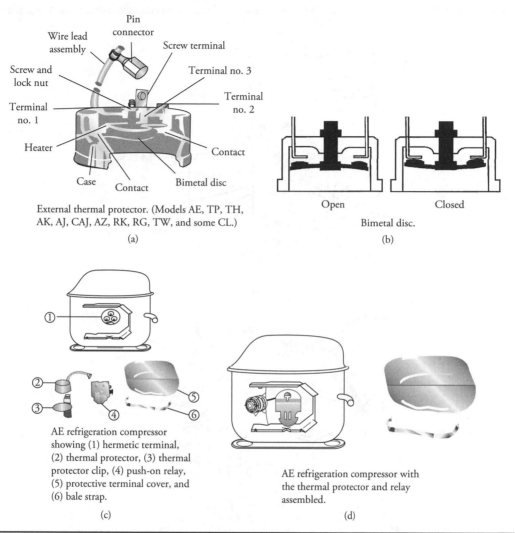

External thermal protector. (Models AE, TP, TH,
AK, AJ, CAJ, AZ, RK, RG, TW, and some CL.)

(a)

Bimetal disc.

(b)

AE refrigeration compressor
showing (1) hermetic terminal,
(2) thermal protector, (3) thermal
protector clip, (4) push-on relay,
(5) protective terminal cover, and
(6) bale strap.

(c)

AE refrigeration compressor with
the thermal protector and relay
assembled.

(d)

FIGURE 5-7 (a) The compressor overload protector; (b) an internal view of the compressor overload
protector; (c) an illustration of the components before they are attached to the exterior of the
compressor; and (d) an illustration of the components after they are attached to the exterior of the
compressor.

non-serviceable part, and it should be replaced with a duplicate of the original. If the
compressor has an internal overload, and it is diagnosed as being defective, the compressor
will have to be replaced by a certified technician.

Condenser Coil

A condenser coil is manufactured out of copper, aluminum, or steel tubing, with metal fins
attached to the tubing to assist in dissipating the heat (Figure 5-8a and 5-8b). The main
purpose of the condenser coil is to dissipate the heat from the refrigerant along with the

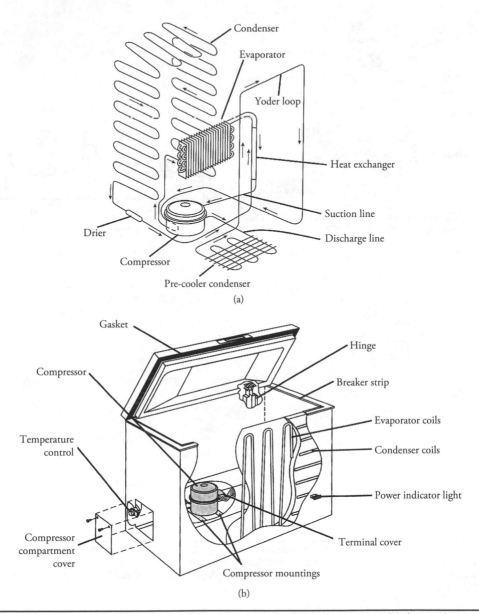

FIGURE 5-8   (a) The condenser coil in a refrigerator's sealed refrigeration system; (b) the condenser coil in a freezer's sealed refrigeration system.

heat from the compression stage of the compressor. Condenser coils in refrigerators, freezers, and air conditioners usually have a fan motor and fan blade circulating the air across the condenser coil to assist in removing the heat from the refrigerant. On some models of refrigerators and freezers, the condenser coils are static-cooled whereby the heat from the condenser coil will rise and dissipate into the room. In some models of freezers, the condenser coil is attached to the inside of the outer cabinet so that the cabinet is the heat transfer medium. In some models of air conditioners the condenser coil is water cooled and does not use a fan motor and fan blade. Some manufacturers have designed their products to be used with a water tower, lake or pond water, or well water as a heat transfer medium.

Evaporator Coil

The evaporator coil, manufactured in the same manner as the condenser coil, absorbs the heat from the food product or from the air in a room to be cooled (Figures 5-8*a* and 5-8*b*, 5-9, and 5-10). Manufacturers design evaporator coils in different shapes and designs to meet the needs of specific products. Most refrigerator models and air conditioners use a fan motor and fan blade to circulate the air across the evaporator coil, which absorbs the heat. Other refrigerator or freezer models have rows of tubing placed between two plates, which are welded together to form an interior cabinet in the freezer compartment or shelving.

Metering Device

The most commonly used metering device in a domestic refrigerator and room air conditioner is a capillary tube, also called a *cap tube* (Figures 5-3 and 5-10). This metering device is manufactured out of copper tubing, with a very small-diameter opening at both ends, and

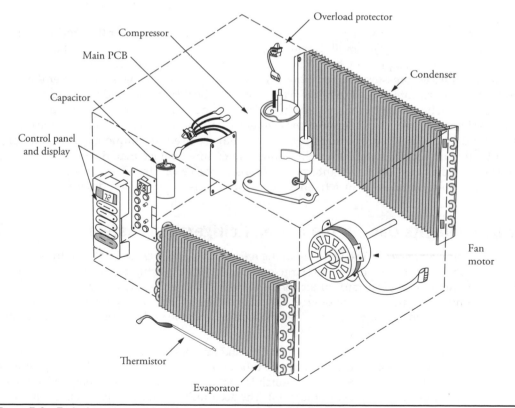

FIGURE 5-9   Typical component locations in a room air conditioner (RAC).

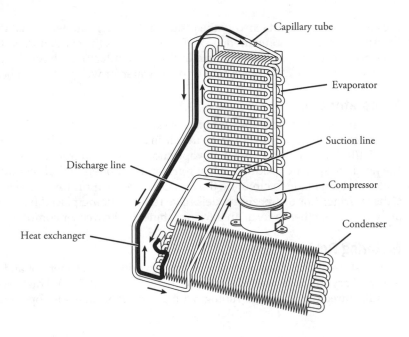

FIGURE 5-10
Typical components including an accumulator and drier of a domestic refrigerator sealed refrigeration system.

is premeasured at a specific length for maximum heat load efficiency for the product being cooled or refrigerated. The small-diameter opening in the cap tube will restrict the flow of refrigerant, causing a pressure drop from the high-pressure side (condenser) of the sealed system to the low-pressure side (evaporator). This will allow the refrigerant to enter the evaporator at the rate needed to remove the unwanted heat from the product being cooled or refrigerated. There are no moving parts in a cap tube, and it does not require any service. The capillary tube is located between the condenser coil and the evaporator coil at the end of the liquid line. Other types of metering devices that are used in commercial refrigeration, residential and commercial air conditioning, and commercial ice machines are thermostatic expansion valve, automatic expansion valve, electronic expansion valve, float type metering device, or a fixed-bore (orifice or piston) metering device.

Other Components Used in the Sealed Refrigeration System

- *Refrigeration drier.* The function of the refrigerant drier (Figure 5-11) is to trap any moisture, contaminants, and large particulate matter, within the sealed refrigeration system. In a refrigerator or air conditioner, the drier is installed within the liquid line before the metering device. The drier can be either a unidirectional flow or bidirectional flow drier. The drier must be compatible with the oils and refrigerants that are used in today's products.

- *Accumulator.* Some models of refrigerators and freezers have accumulators (Figure 5-11). An accumulator is a device located at the end of the evaporator coil. Its appearance will look as if the tubing size is much larger. It looks like a small elongated storage tank attached to the evaporator coil. The accumulator allows liquid refrigerant to be collected in the bottom of the accumulator and stays there. Only vapor refrigerant will be drawn from the accumulator and returned to the compressor.

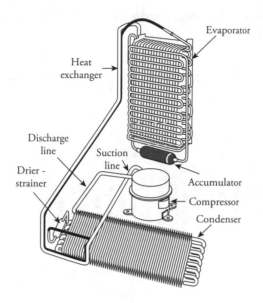

FIGURE 5-11
Typical components including an accumulator and drier of a domestic refrigerator sealed refrigeration system.

On air conditioners, the accumulator is located on the suction line returning to the compressor. The compressor and accumulator are one component.

- *Heat exchanger.*
  The heat exchanger (Figure 5-10) consists of the suction line soldered to the capillary tube. The reason for this is to make sure that there is good contact between the two lines. These two tubes are located inside a refrigerator or freezer. There are two purposes for the heat exchanger:
  - The capillary tube will give up some of its heat to the cooler suction line, thus allowing the refrigeration system to run more efficiently.
  - No liquid refrigerant will be allowed to return to the compressor. The warmer refrigerant in the capillary tube will boil off and vaporize any remaining liquid refrigerant in the suction line.

Diagnosing Sealed Systems

While servicing the refrigerator, freezer, or air conditioner, do not overlook the simple things that might be causing the problem. Before you begin testing and servicing a refrigeration product or air conditioner, check for the most common faults. Inspect all wiring connections for broken or loose wires. Check and see if all components are running at the proper times during the cycle. Test the temperatures in the refrigerator, freezer cabinet, or air conditioner inlet and outlet grills.

The following tables (Tables 5-4 and 5-5) are intended to serve as aids to assist the technician in determining if there is a sealed system malfunction, providing that all other components are functioning correctly within the product or conditions that mimic a sealed system failure. You must rule out everything else before you enter the sealed system.

NOTE *Before you enter a sealed system, you must be certified to do so.*

Refrigeration Sealed System Diagnosis Chart

Sealed System Condition	Suction Line Pressure (Technicians Only)	Liquid Line Pressure (Technicians Only)	Suction Line to Compressor	Compressor Discharge Line	Condenser Coil	Capillary Tube	Evaporator Coil	Frost Line	Amperage or Wattage	Pressure Equalization Rate
Normal Operation	Normal pressure readings	Normal pressure readings	Slightly below room temperature	**WARNING** Very hot to the touch	**WARNING** Very hot to the touch	Warm	Cold coil and maintaining proper temperatures	Suction line from inside box not frozen	Normal meter readings	Normal
Sealed System Overcharged	Higher than normal pressure readings	Higher than normal pressure readings	Heavily frosted; may be very cold to the touch	Between slightly warm to hot	Between hot to warm to the touch	Cool; below room temperature	Cold coil and possibly not maintaining temperatures	All the way back to the suction line	Higher than normal meter readings	Normal to slightly longer
Sealed System Undercharged	Pressure readings are lower than normal	Pressure readings are lower than normal reading	Warm to the touch; possibly near room temperature	**WARNING** Hot to the touch	The entire coil feels warm	Warm	Inlet to coil feels extremely cold while the outlet from the coil will be below room temperature	Partial	Lower than normal meter readings	Normal
Partial Restriction Within Sealed System	Lower than normal pressure readings, possibly in a vacuum	Intermittent lower than normal reading	Warm to the touch; possibly near room temperature	**WARNING** Very hot to the touch	The top passes in the coil are warm and the lower passes are cool, near to room temperature	Feels like room temperature between cool to colder	Inlet to coil feels extremely cold while the outlet from the coil will be below room temperature	Intermittent / Frost line will begin to grow in length	Lower than normal meter readings	Intermittent
Complete Restriction Within Sealed System	Pressure readings are in a deep vacuum	Ambient readings	Feels the same as room temperature	Feels the same as room temperature	Feels the same as room temperature	Feels the same as room temperature	No refrigeration or air conditioning	None	Lower than normal meter readings	No equalization
Out of Refrigerant Possible Leak in System	The pressure reading will be from 0 PSIG to 30" vacuum	Atmospheric reading	Feels the same as room temperature	Can feel like cool to hot	Feels the same as room temperature	Feels the same as room temperature	No refrigeration or air conditioning	Non existent	Lower than normal meter readings	Normal
Low Capacity Compressor	Higher than normal pressure readings	Lower than normal readings	Cool to room temperature	Cooler than normal	Low	Warm to box temperature	Partial or half of the evaporator frost pattern	Partial to non-existent	Lower than normal	Quicker than normal

TABLE 5-4 Refrigeration Sealed-System Diagnosis Chart

Conditions That Mimic Sealed System Failures

Conditions	Amperage or Wattage	Condenser Coil Temperature	Frost Line	Compressor Discharge Line Temperature	Low Side Pressure (for Service Technicians Only)	High Side Pressure (for Service Technicians Only)	Fresh Food Compartment Temperature	Freezer Compartment Temperature
Plugged Condenser Coil	Higher than normal	Higher than normal	Full	Higher than normal	Higher than normal	Higher than normal	Warmer than normal readings	Warmer than normal readings
Blocked Condenser Fan Assembly	Higher than normal	Higher than normal	Full	Higher than normal	Higher than normal	Higher than normal	Warmer than normal readings	Warmer than normal readings
Blocked Evaporator Fan Assembly	Lower than normal	Lower than normal	Frost back to compressor	Lower than normal	Lower than normal	Lower than normal	Warmer than normal readings	Warmer than normal readings
Evaporator Coil Iced Up (Defrost Failure)	Lower than normal	Lower than normal	Frost back to compressor	Lower than normal	Lower than normal	Lower than normal	Warmer than normal readings	Warmer than normal readings
High Head Load	Higher than normal	Higher than normal	Full	Higher than normal	Higher than normal	Higher than normal	Warmer than normal readings	Warmer than normal readings
High Ambients	Higher than normal	Higher than normal	Full	Higher than normal	Higher than normal	Higher than normal	Warmer than normal readings	Warmer than normal readings
Damper Failed Closed	Lower than normal	Lower than normal	Full	Lower than normal	Lower than normal	Lower than normal	Warmer than normal readings	Cooler than normal readings
Damper Failed Open	Slightly higher than normal	Slightly higher	Full	Normal	Slightly higher than normal	Normal	Cooler than normal	Normal to slightly warmer readings

TABLE 5-5 Conditions That Will Mimic a Sealed System Failure

Basic Electricity and Electronics

A technician must be knowledgeable in electrical theory to be able to perform maintenance and to diagnose and repair air conditioning, heating, and refrigeration equipment properly. Although this chapter cannot cover all there is to know about electricity and electronics, it will provide the basics. In the field of air conditioning, heating, and refrigeration, the greatest number of potential problems is in the electrical portions of the product.

Electrical Wiring

The flow of electricity from a power source to the home or business can be made easier to understand by comparing it to a road map. Electricity flows from a power source to a load. This is similar to a major highway that runs from one location to another.

High-voltage transformers are used to increase voltages for transmission over long distances. The power lines that go to different neighborhoods are like the smaller roads that turn off the major highway. The electricity then goes to a transformer that reduces the voltage going into the home or business. This is the intersection between the small roads and the medium-sized highways. The small road that goes into the neighborhood, and all the local streets are like the wiring that goes inside the home or business.

When all the streets, roads, and highways are connected together the city is accessible. This is similar to having electricity flowing from the power source to all of the outlets in the home or business.

Imagine driving down a road and coming to a drawbridge (in this case, the switch), and it opens up. This stops the flow of traffic (electricity). In order for traffic (or electricity) to flow again, the drawbridge must close.

What Is a Circuit?

A *circuit* is a complete path through which electricity can flow and then return to the power source. Figure 6-1 is an example of a complete circuit. To have a complete path (or *closed circuit*), the electricity must flow from Point A to Point B without interruption.

FIGURE 6-1
The complete circuit. Current flows from Point A, through the light bulb, and then back to Point B.

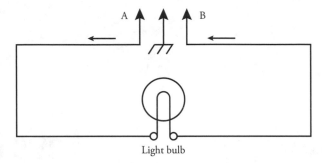

Light bulb

FIGURE 6-2
With the switch open, the current flow is interrupted.

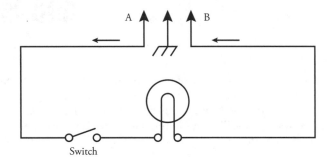

Switch

When there is a break in the circuit, the circuit is *open*. For example, a break in a circuit is when a switch is turned to its "off" position. This will interrupt the flow of electricity, or current, as shown in Figure 6-2. When a broken circuit is suspected, it is necessary to discover the location of the opening.

Circuit Components

In air conditioning, heating, and refrigeration equipment, an electric circuit has four important components:

- *Power source.* This source might be a battery or the electricity coming from the wall outlet or circuit breaker panel. Without the applied voltage, current cannot flow.

- *Conductors.* A conductor will usually be a wire and sometimes the metal chassis (frame). The function of the wire conductor is to connect a voltage source to a load.

- *Loads.* These are the components that do the actual work in the air conditioning, heating, or refrigeration equipment. A load is anything that uses up some of the electricity flowing through the circuit. For example, motors turn the belt, which turns the pulley. That, in turn, turns the gear box and auger in a flaker ice machine. Other examples are heating elements and solenoids.

- *Controls.* These control the flow of electricity to the loads. A control is a switch that is either manually operated by the user of the product or operated by the air conditioning, heating, or refrigeration equipment itself.

Three Kinds of Circuits

You will come across three kinds of circuits: *series circuits, parallel circuits,* and *series-parallel circuits* (a combination of series and parallel circuits).

Series Circuits

The components of a *series circuit* are joined together in successive order, each with an end joined to the end of the next (Figure 6-3). There is only one path that electricity can follow. If a break occurs anywhere in the circuit, the electricity, or current flow, will be interrupted and the circuit will not function (Figure 6-4). Figure 6-5 shows some of the many different shapes of series circuits, all of which are used in wiring diagrams. In each series circuit, there is only one path that electricity can follow. There are no branches in these circuits where current can flow to take another path. Electricity only follows one path in a series circuit.

Parallel Circuits

The components of a *parallel circuit* are connected across one voltage source (Figure 6-6). The voltage to each of these branches is the same. The current will also flow through all the branches at the same time. The amount of current that will flow through each branch is determined by the load, or resistance, in that branch.

Figures 6-7 and 6-8 show examples of parallel circuits. If any branch has a break in it, the current flow will only be interrupted in that branch. The rest of the circuits will continue to function.

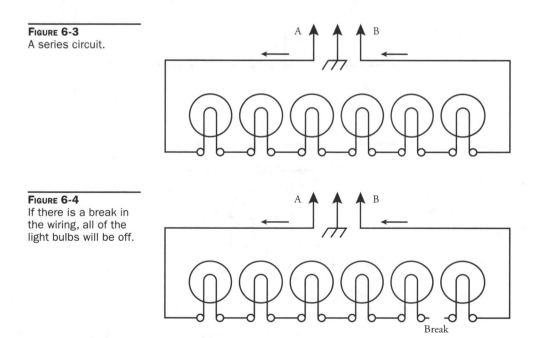

FIGURE 6-3
A series circuit.

FIGURE 6-4
If there is a break in the wiring, all of the light bulbs will be off.

Break

FIGURE 6-5
Series circuits in all sorts of shapes in wiring diagrams.

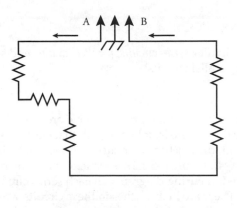
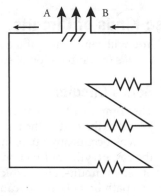
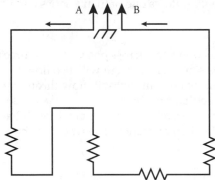
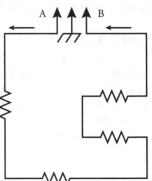

FIGURE 6-6
Parallel circuit.

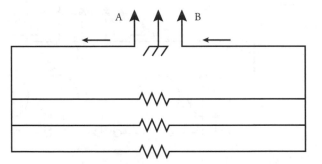

FIGURE 6-7
Another parallel circuit.

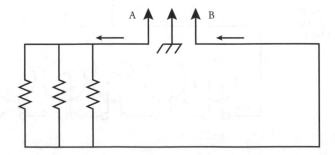

FIGURE 6-8
Notice that as in series circuits, the same parallel circuit can be drawn in many different ways.

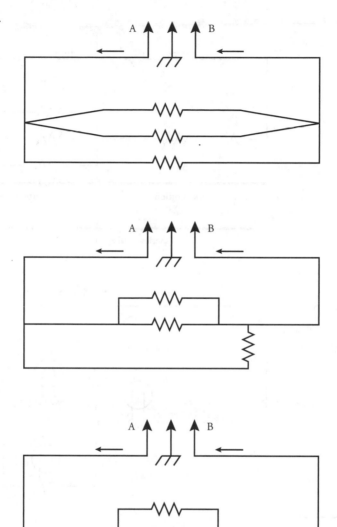

Series-Parallel Circuits

A *series-parallel circuit* is a combination of series circuits and parallel circuits. In many circuits, some components are connected in series to have the same current, but others are in parallel for the same voltage (Figure 6-9). This type of circuit is used where it is necessary to provide different amounts of current and voltage from the main source of electricity that is supplied to that product.

Series and parallel rules apply to this type of circuit. For example, if there is a break in the series portion of the circuit (Figure 6-10), the current flow will be interrupted for the entire circuit. If the break is in the parallel portion of the circuit (Figure 6-11), the current will be interrupted for only that branch of the circuit. The rest of the circuits will still function.

Figure 6-9
Series-parallel circuit.

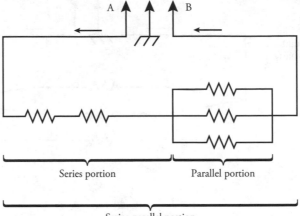

Series portion   Parallel portion

Series-parallel portion

Figure 6-10
Series-parallel circuit
with open in the
series portion of the
circuit. No current flow
in the entire circuit.

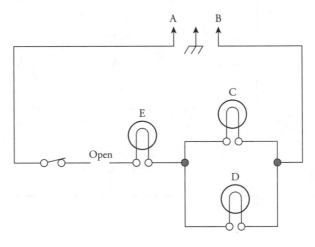

Figure 6-11
Series-parallel circuit
with open in the
parallel portion of the
circuit. Current flows
through D, E, and F,
but not through G.

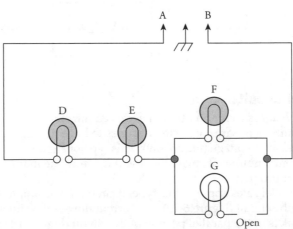

Types of Shorts in a Circuit

When a short occurs in a series circuit, it means there is a zero resistance across an electrical source. Figure 6-12 shows a wire across load C. This short, which is the path of least resistance, allows the current to flow through the wire instead of the load.

A short can be added to part of the series circuit, to bypass one or more of the loads in the circuit. This type of short is referred to as a *shunt*, manufacturers could design this shunt into a circuit. The difference between a short and a shunt is as follows:

- With a short there is no resistance to the flow of current in a circuit.

- A shunt will form a bypass around the load(s), but the circuit will still offer resistance to the flow of current.

In Figure 6-13, if a short is connected in the series circuit between C and D, the loads E, F, and G would be bypassed because they are shorted out of the circuit. Electricity takes the path of least resistance.

In Figure 6-14, if a permanent shunt is connected in the series circuit between points C and D, loads F and G would be bypassed because they were shorted out of the circuit.

In Figure 6-15, if the manufacturer designed a load to be turned on and off, they will install a switch to act as a shunt across a load. When switch X is closed, load G is turned off, allowing loads E and F to remain on.

In Figure 6-16, the parallel circuit will allow current to flow from point A through the loads, C and D, and back to point B.

FIGURE 6-12
A wiring short in a series circuit.

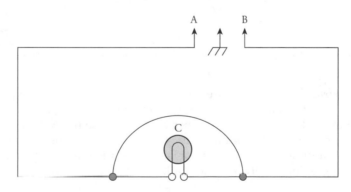

FIGURE 6-13
A series circuit with a short across loads E, F, and G.

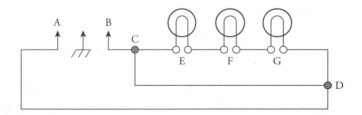

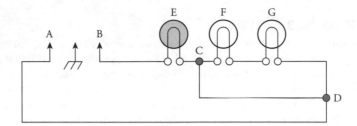

FIGURE 6-14
A series circuit with a permanent shunt across loads F and G.

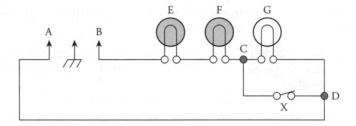

FIGURE 6-15
A series circuit with a switch acting as a temporary shunt across load G.

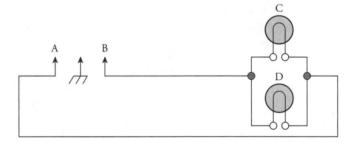

FIGURE 6-16
A parallel circuit with two loads.

In the parallel circuit in Figure 6-17, a short was placed between points X and Y, shorting out loads C and D. The current will flow from point A through the loads and through the shorted wire back to point B. Remember, electricity take the path of least resistance.

In the series-parallel circuit in Figure 6-18, the current will flow from point A, through the short beginning at point C, to point D, and back to point B, bypassing loads E, F, G, and H. Keep in mind, electricity takes the path of least resistance.

FIGURE 6-17
A parallel circuit with a short between loads C and D.

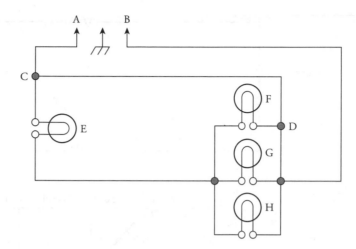

FIGURE 6-18
A series-parallel circuit with a short.

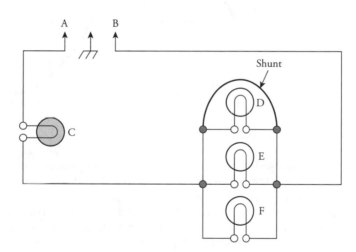

FIGURE 6-19
A series-parallel circuit with a shunt.

In Figure 6-19, current will flow through the series portion of the circuit from point A through the load C. In the parallel portion of the circuit, current will flow from load C through the wire (shunt) across load D, and back through point B. The three loads would be bypassed. Electricity takes the path of least resistance.

Types of Electric Current

There are two types of electric current:

- *Direct current* (DC) flows continuously in the same direction (Figure 6-20).
- *Alternating current* (AC) flows in one direction and then reverses itself to flow in the opposite direction along the same wire. This change in direction occurs 60 times per second, which equals 60 Hz (Figure 6-21).

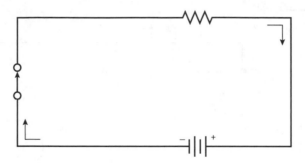

FIGURE 6-20
A simple DC electrical circuit. Current flows from the negative side of the battery through the switch and load, and back to the positive side of the battery.

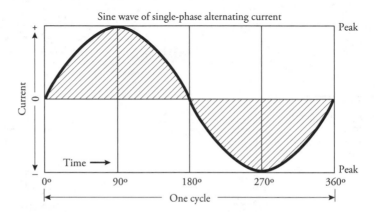

FIGURE 6-21
A waveform of a single-phase alternating current.

Sine wave of single-phase alternating current

Peak

Current

Time ⟶

Peak

0°   90°   180°   270°   360°

One cycle

Direct current is used in automobile lighting, flashlights, and cordless electric appliances (such as toothbrushes, shavers, and drills), and in some air conditioning, heating, and refrigeration equipment.

Alternating current is used in most homes and businesses. This current can be transmitted more economically over long distances than direct current can. Alternating current can also be easily transformed to higher or lower voltages.

Theory of Current Flow

When servicing electronic solid-state circuits in air conditioning, heating, or refrigeration products, the technician should be aware that explanation of the circuit is often given assuming conventional current flow as opposed to electron flow. Conventional current flow theory states that current flow is from positive to negative. Electron current flow theory states that current flows from negative to positive.

Ohm's Law

Ohm's law states (I = E/R); the current which flows in a circuit is directly proportional to the applied voltage and inversely proportional to the resistance. So, an increase in the voltage will increase the current as long as the resistance is held constant. Alternately, if the resistance in a circuit is increased and the voltage does not change, the current will decrease.

The second version states (E = I × R); it can be seen from this equation that if either the current or the resistance is increased in the circuit (while the other is unchanged), the voltage will also have to increase.

The third version states (R = E/I); if the current is held constant, an increase in voltage will result in an increase in resistance. Alternately, an increase in current while holding the voltage constant will result in a decrease in resistance.

In any of the three formulas, when two elements of the electric circuit are known, the unknown factor can be calculated.

Ohms

Resistance is measured in *ohms* and opposes the flow of electrons (current). An instrument that measures resistance is known as an ohmmeter. Figure 6-22 is a schematic showing an ohmmeter connected to read the resistance of R1. The resistance of any material depends on its type, size, and temperature. Even the best conductor offers some opposition to the flow of electrons. Figure 6-23 shows another type of meter, the digital multimeter, used for measuring ohms. The fundamental law used to find resistance is stated as follows: The resistance (*R*) in ohms is equal to the potential difference measured in volts (V), divided by the current in amperes (A). The equation is: *R* = V/A.

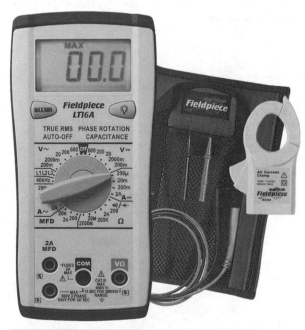

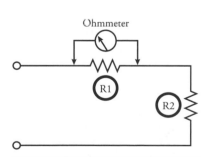

FIGURE 6-22 An ohmmeter connected to read resistance.

FIGURE 6-23 A digital multimeter for measuring ohms, volts, and capacitance.

Amperes

Current is measured in *amperes*. The term ampere refers to the number of electrons passing a given point in 1 second. When the electrons are moving, there is current. The ammeter is calibrated in amperes, which we use to check for the amount of current in a circuit.

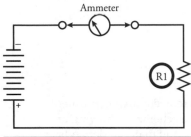

Ammeter

An instrument that measures amperes is known as an ammeter. Figure 6-24 is a schematic showing an ammeter connected in a circuit to measure the current in amperes. Figure 6-25 shows an ammeter that is used in diagnosing electrical problems with air conditioning, heating, or refrigeration equipment. Current is the factor that does the work in the circuit (light the light, ring the buzzer). The fundamental law to find current is stated as follows: The current in amperes (A) is equal to the potential difference measured in volts (V), divided by the resistance in ohms (R). The equation is $A = V/R$.

FIGURE 6-24   An ammeter connected in a circuit, measuring amperes.

Volts

Electromotive force is measured in *volts*. This is the amount of potential difference between two points in a circuit. It is this difference of potential that forces current to flow in a circuit. Potential difference of 1 V is the electromotive force required to force 1 A of current through 1 Ω of resistance.

FIGURE 6-25
In the ammeter, the jaws clamp around a wire to measure the amperage of a circuit.

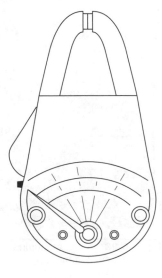

An instrument that measures voltage is known as a voltmeter. Figure 6-26 is a schematic showing a voltmeter connected in the circuit to measure the voltage. Voltmeter 1 is connected to read the applied (or source) voltage. Voltmeter 2 is connected to measure the voltage drop, or potential difference, across R2. Figure 6-27 shows an actual volt-ohm-milliameter (VOM) that is used in measuring voltage.

The fundamental law to find voltage is stated as follows: The potential difference measured in volts (V) is equal to the current in amperes (A) multiplied by the resistance in ohms (R). The equation is $V = A \times R$.

FIGURE 6-26
A voltmeter connected in a circuit to measure voltage.

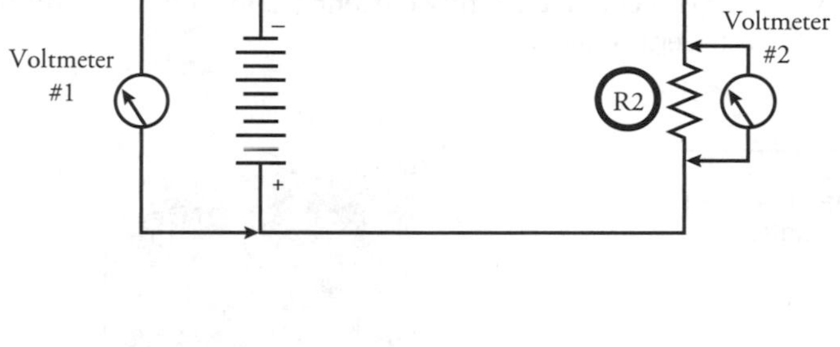

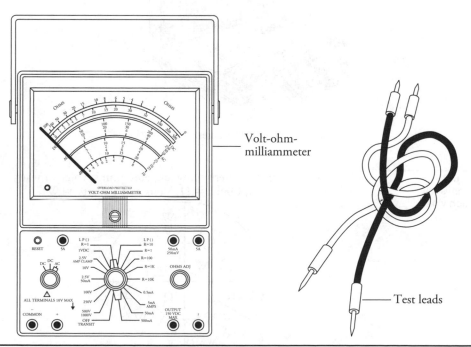

FIGURE 6-27   The volt-ohm-milliammeter with test leads.

Watts

Power is measured in *watts*, and an instrument that measures watts is known as a wattmeter (Figure 6-28). One watt of power equals the work done in 1 second, by 1 V of potential difference, in moving 1 C (coulomb) of charge. One C/second (coulomb per second) is equal to 1 A. Therefore, the power in watts (W) equals the product of amperes (A) times volts (V). The equation is $W = A \times V$.

Ohm's Law Equation Wheel

The *Ohm's law equation wheel* in Figure 6-29 shows the equations for calculating any one of the basic factors of electricity. Figure 6-30 shows the cross-reference chart of formulas as used in this text. If you know any two of the factors (V = voltage, A = amperage, R = resistance, W = power), you can calculate a third. To obtain any value in the center of the equation wheel for direct or alternating current, perform the operation indicated in one segment of the adjacent outer circle.

Figure 6-28
The wattmeter is used to measure watts.

Digital wattmeter

Figure 6-29
The Ohm's law equation wheel.

Conversion chart for determining amperes, ohms, volts, or watts
(Amperes = A, Ohms = Ω, Volts = V, Watts = W)

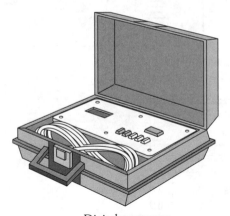

FIGURE 6-30
The cross-reference
chart of formulas.

Term	Measured in	Referred to in formulas as	Identification used in this text
Amperage	Amperes (Amps)	I	A for amps
Current	Amperes (Amps)	I	A for amps
Resistance	Ohms	Ω or R	R for resistance
Voltage	Volts	V or E	V for volts
Electromotive force	Volts	V or E	V for volts
Power	Watts	W	W for watts

Example 6-1: A 2400-W heating element is connected to a 240-V circuit. How many amps does it draw?

When finding amperage, the formula will be found in the Amperes section of the wheel.

$$\frac{W(watts)}{V(volts)} = A(amps)$$

Then, solving for amperage:

$$\frac{2,400 \text{ W}}{240 \text{ V}} = 10 \text{ A}$$

What is the resistance (ohms)?

$$\frac{V^2(volts\ squared)}{W(watts)} = ohms(\Omega)$$

Then, solving for resistance:

$$\frac{2,402\ (volts\ squared)}{2,400\ (W)} = 24\ \Omega$$

Example 6-2: What is the resistance of a 100-W light bulb if the voltage is 120 V and the current is 0.83 A?

When finding resistance, the formula will be found in the Ohms section of the wheel.

$$\frac{V\ (volts)}{A\ (amps)} = R\ (resistance)$$

$$\frac{120\ V}{0.83\ A} = R\ (resistance)$$

$$145\ \Omega\ (resistance)$$

Example 6-3: What is the voltage of a circuit if the resistance of the load is 48 Ω, and current is 5 A?

$$amps\ (A) \times ohms\ (R) = volts\ (V)$$

$$5\ A \times 48\ \Omega = 240\ V$$

Wiring Diagram Symbols

These symbols are commonly used in most wiring diagrams. Study each symbol so that you can identify it by sight (Tables 6-1 to 6-4).

Terminal Codes

Terminal codes are found on all wiring diagrams. To help you identify the color codes, they are listed in Table 6-5.

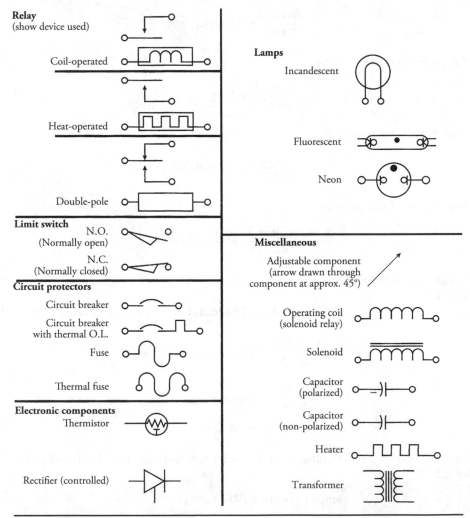

TABLE 6-1 Wiring Diagram Symbols

Temperature-actuated components

(Note: Symbols shown to be used for thermosstats, bimetal switches, overload protectors, or other similar components, as required)

Temp. actuated
(close on heat rise)

Temp. actuated
(open on heat rise)

S.P.S.T.
(open on heat rise)

S.P.D.T.

S.P.D.T.

S.P.S.T.
(two contacts)

S.P.S.T. (adj.)
(close on heat rise)

S.P.D.T. (adj.)

S.P.S.T. (adj.)
(open on heat rise)

S.P.D.T. (adj.)
(with aux. "off" contacts)
(typical example)

S.P.S.T. (with internal heater)
(close on heat rise)

S.P.S.T. (with internal heater)
(open on heat rise)

Combination devices

Relay-magnetic
(arrangement of
contacts as necessary
to show operation)

Relay-thermal
(arrangement of
contacts as necessary
to show operation)

Timer (defrost)

Manual and mechanical switches

Normally closed (S.P.S.T.)
(single-pole, single-throw)

Normally open (S.P.S.T.)
(single-pole, single-throw)

Transfer (S.P.S.T.)
(single-pole, double-throw)

Multi position

Number of terminals

Timer switch

Automatic switch

N.O.
(normally open)

N.C.
(normally closed)

Integral switch
(timer, clock, etc.)

**Push button switch
(momentary or spring seturn)**

Circuit closing
N.O. (normally open)

Circuit opening
N.C. (normally closed)

Two circuit

SPDT
(single-pole, double-throw)

TABLE 6-2   Wiring Diagram Symbols

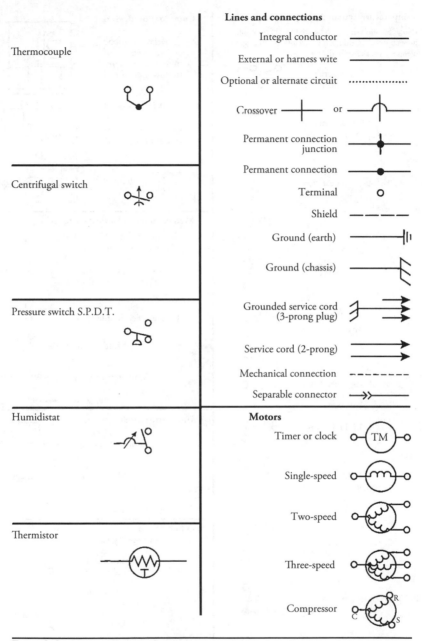

Lines and connections

Integral conductor	
External or harness wite	
Optional or alternate circuit	
Crossover	or
Permanent connection junction	
Permanent connection	
Terminal	○
Shield	
Ground (earth)	
Ground (chassis)	
Grounded service cord (3-prong plug)	
Service cord (2-prong)	
Mechanical connection	
Separable connector	

Motors

Timer or clock	TM
Single-speed	
Two-speed	
Three-speed	
Compressor	

Thermocouple

Centrifugal switch

Pressure switch S.P.D.T.

Humidistat

Thermistor

TABLE 6-3   Wiring Diagram Symbols

Electronic wiring symbols

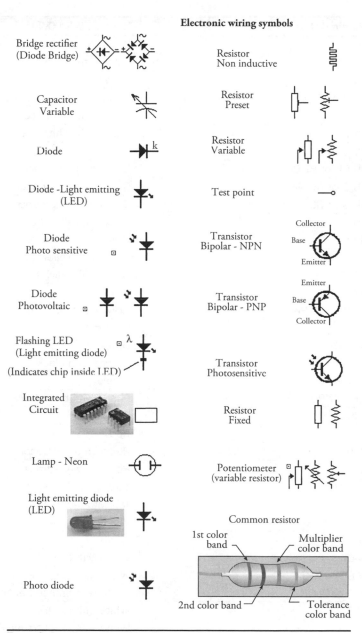

Bridge rectifier (Diode Bridge)

Capacitor Variable

Diode

Diode -Light emitting (LED)

Diode Photo sensitive

Diode Photovoltaic

Flashing LED (Light emitting diode)

(Indicates chip inside LED)

Integrated Circuit

Lamp - Neon

Light emitting diode (LED)

Photo diode

Resistor Non inductive

Resistor Preset

Resistor Variable

Test point

Transistor Bipolar - NPN
Collector
Base
Emitter

Transistor Bipolar - PNP
Emitter
Base
Collector

Transistor Photosensitive

Resistor Fixed

Potentiometer (variable resistor)

Common resistor

1st color band

Multiplier color band

2nd color band

Tolerance color band

TABLE 6-4   Electronic Wiring Diagram Symbols

Terminal Color Code	Harness Wire Color
BK	Black
BK-Y	Black with yellow tracer
BR	Brown
BR-O or BR-OR	Brown with orange tracer
BR-R	Brown with red tracer
BR-W	Brown with white tracer
BU or BL	Blue
BU-BK or BL-BK	Blue with black tracer
BU-G or BU-GN	Blue with green tracer
BU-O or BU-OR	Blue with orange tracer
BU-Y	Blue with yellow tracer
G or GN	Green
B-Y or GN-Y	Green with yellow tracer
G-BK	Green with black tracer
GY	Gray
GY-P or GY-PK	Gray with pink tracer
LBU	Light blue
O or OR	Orange
O-BK or OR-BK	Orange with black tracer
P or PUR	Purple
P-BK or PUR-BK	Purple with black tracer
P or PK	Pink
R	Red
R-BK	Red with black tracer
R-W	Red with white tracer
T or TN	Tan
T-R	Tan with red tracer
V	Violet
W	White
W-BK	White with black tracer
W-BL or W-BU	White with blue tracer
W-O or W-OR	White with orange tracer
W-R	White with red tracer
W-V	White with violet tracer
W-Y	White with yellow tracer
Y	Yellow
Y-BK	Yellow with black tracer
Y-G or Y-GN	Yellow with green tracer
Y-R	Yellow with red tracer

TABLE 6-5　Terminal Codes

Wiring Diagrams

In this section, each of the three examples presented will take you through a step-by-step process in how to read wiring diagrams.

Example 6-4: Take a look at a simple wiring diagram for a refrigerator (Figure 6-31). Note the black wire on the diagram. This is the wire that goes to the temperature control. The circuit is not energized when the temperature control is in the "off" position. When the temperature control knob is turned to the "on" position (switch contacts closed) and the circuit is energized, current will flow through the temperature control, through the red wire, through the overload protector, through the compressor and the relay, and back through the white wire to the line cord.

Example 6-5: The wiring diagram in Figure 6-32 is for a refrigerator. Assume that the thermostat is calling for cooling and the compressor is running. With your finger, trace the active circuits. The thermostat in the wiring diagram for the refrigerator is closed. The evaporator fan motor and the condenser fan motor are running. Voltage is supplied through the overload to the relay. Current is flowing through the relay coil to the compressor-run-winding. Also notice that the door switch is open and the refrigerator light is off. When the temperature in the refrigerator satisfies the thermostat, the thermostat switch contacts will open, thus turning off the compressor, the evaporator fan motor, and the condenser fan motor.

Example 6-6: The wiring diagram in Figure 6-33 is for a no-frost refrigerator. Note the defrost timer in the lower-left part of the diagram. The defrost timer switch contact is closed to contact 4, the thermostat is calling for cooling, and the compressor is running. Trace the active circuits with your finger. Voltage is supplied to the defrost timer terminal 1. Current will flow through the defrost 2 timer motor to the white wire and then back to the line cord. At the same time, current flows through contact 4 in the defrost timer to the thermostat. At this point, the current passes through the thermostat to a junction and splits in two directions. Current will flow through the temperature control, through the overload protector, through the compressor and the relay, and back through the white wire to the line cord. At the same time, the compressor is running and current will flow through the evaporator and condenser fan motors.

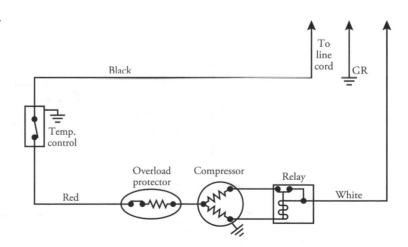

FIGURE 6-31
A simple wiring schematic of a refrigerator circuit.

Figure 6-32
The wiring schematic of a refrigerator circuit.

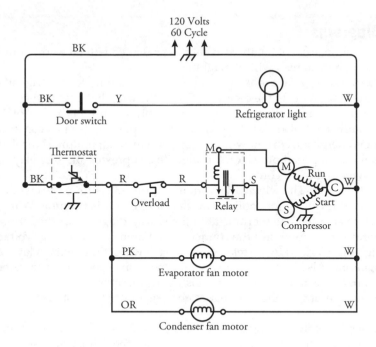

Figure 6-33
The wiring schematic of a no-frost refrigerator circuit.

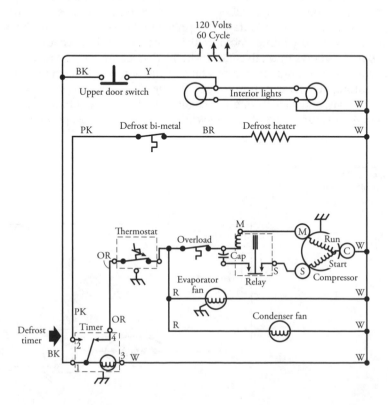

When the defrost timer activates the defrost cycle, the defrost timer switch contact is closed to contact 2, and the compressor is not running. The evaporator and condenser fan motors will also stop running. The current will flow from terminal 2, through the defrost bi-metal, through the defrost heater, and back through the white wire to the line cord. With your finger, trace the active circuits.

Sample Wiring Diagrams

The diagrams in Figures 6-34 and 6-35 illustrate a pictorial diagram and a schematic diagram. The pictorial diagram shows the actual picture of the components, and the schematic diagram uses symbols for the components. Figure 6-36 shows a pictorial diagram of a refrigerator. In this type of diagram, you can see where the components are actually located.

Multimeters

A *multimeter* in Figure 6-37 is a diagnostic tool used by technicians to troubleshoot electrical circuits and components in air conditioning, heating, and refrigeration equipment. The multimeter is a tool that combines the functions of a voltmeter, ammeter, and ohmmeter (VOM). This meter measures voltage, resistance, current (amps) as well as some other functions as well, into a single testing instrument. The technician can choose between an analog and a digital liquid crystal display (LCD) on the meter. Digital multimeters (DMM) are more commonly used today.

Multimeter Safety Information

Safety starts with accident prevention. Injuries are usually caused when an individual trying to use a multimeter fails to read how to use the meter. Listed below are some tips to help the homeowner, business owner, maintenance engineer, and the HVAC/R technician to correctly and safely operate a multimeter to perform maintenance on air conditioning, heating, and refrigeration equipment.

Any person who cannot use a multimeter should *not* attempt to install, maintain, or repair any air conditioning, heating, or refrigeration equipment. Any improper use of the multimeter will create a risk of personal injury, as well as property damage.

More safety tips:

- Make sure the test leads and the function switch are in the correct position for the desired measurement.
- If the meter and/or test leads are damaged, do not use the meter.
- Never test a circuit for resistance with the electricity turned on.
- Never use a meter on circuits that exceed the meters specifications.
- Remember to be careful when working with voltages above 60 VDC or 30 VAC root mean square (rms). Such voltages pose a shock hazard.
- When using the test leads, keep your fingers behind the finger guards when making measurements.
- To avoid false readings, which could lead to possible electric shock or personal injury, replace the batteries when the battery indicator appears in the screen.

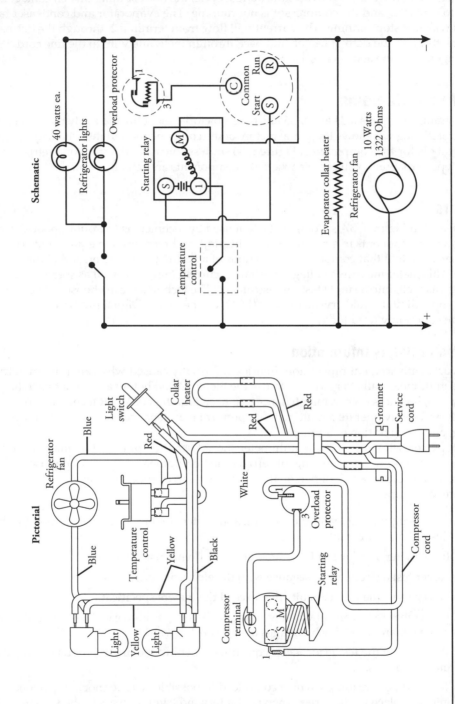

FIGURE 6-34 A pictorial diagram and a schematic diagram. Both show the same components.

FIGURE 6-35
A pictorial diagram
and schematic
diagram.

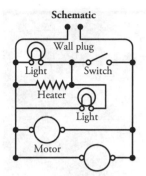

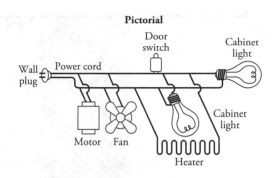

Schematic

Pictorial

FIGURE 6-36
A pictorial diagram of
a refrigerator, showing
where the
components are
located.

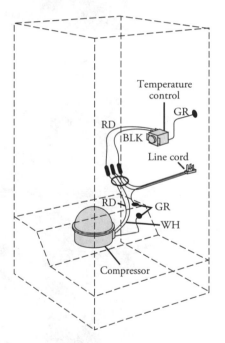

WIRE CODE	
COLOR	CODE
RED	RD
WHITE	WH
BLACK	BLK
ORANGE	OR
GREEN	GR
BLUE	BLU

- There are many different types and brands of VOMs (Figure 6-38). However, all VOMs are used in the same way to measure voltage, current, or resistance. Most VOMs will have the following:
  - *Test leads.* These are the wires coming from the meter to the part being tested.
  - *Meter scales and pointer (or a digital display on a digital meter).* These show the amount of whatever value you are measuring.
  - *Function switch.* This allows you to select whether you will be measuring AC or DC voltage (volts), current (amps), or resistance (ohms).
  - *Range selector knob.* This allows you to select the range of values to be measured. On many meters, you can select functions and ranges with the one switch (as in the meter pictured in Figure 6-38a and 6-38c).

FIGURE **6-37**
(*a*) Digital multimeter
and (*b*) digital clamp-
on multimeter.

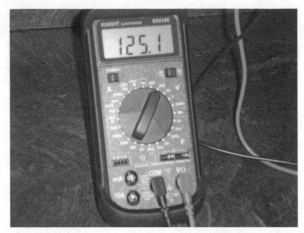

(a)

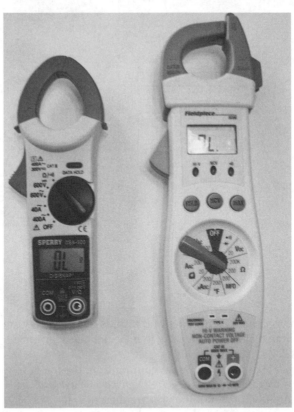

(b)

FIGURE 6-38
(a) Volt-ohm-milliammeter;
(b) VOM and ammeter multimeter combination; and (c) a digital multimeter.

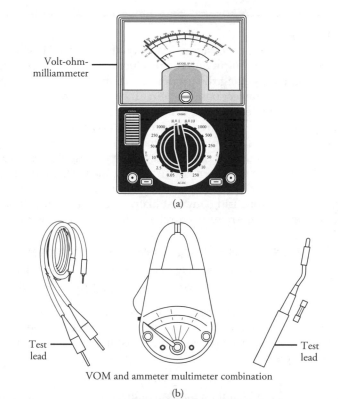

Volt-ohm-milliammeter

(a)

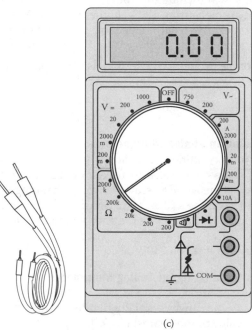

Test lead

Test lead

VOM and ammeter multimeter combination

(b)

(c)

Measuring Voltage

If you don't have the proper voltage as stated on the data tag attached to the air conditioning, heating, or refrigeration equipment, the product won't function properly. You can find out whether the product is getting the proper voltage by measuring the voltage at the wall outlet (receptacle), disconnect panel, or circuit breaker panel. If the air conditioning, heating, or refrigeration equipment isn't getting the proper voltage, nothing else you do to fix it will help. Voltage is measured between two points making sure to have the test probes making good contact at each point being measured. By using the three-point test method to test live voltage in a circuit, the technician will be placing the test probes on a known live circuit first, then test the unknown circuit, finally, the technician will place the test probes back on the live circuit to test the known live circuit again.

For your safety, before using any test instrument, it is your responsibility as a technician to read and understand the manufacturer's instructions on how the test instrument operates.

Making the Measurement Using an Analog Multimeter

Making voltage measurements is easy once you know how to select and read the scales on your meter. When measuring voltage, you should perform the following steps:

- Attach the probes (another term for *test leads*) to the meter. Plug the black probe into the meter jack marked negative or common. Plug the red probe into the positive outlet. We remove the probes (test leads) in reverse order, red probe first and then the black probe.
- Set the range selector knob switch to AC VOLTS.
- Select a range that will include the voltage you are about to measure (higher than 125 V AC if you are measuring 120 V; higher than 230 V AC if you are measuring 220 V.) If you don't know what voltage to expect, use the highest range, and then switch to a lower range if the voltage is within that amount.
- Touch the tips of the probes to the terminals of the part to be measured.
- Read the scale.
- Decide what the reading means.

Selecting the Scales Using an Analog Multimeter

To measure the voltage in air conditioning, heating, or refrigeration equipment, use the AC/DC scales (Figure 6-39). These same scales are used for both AC and DC readings. The numbers on the right side of the AC/DC scales tell you what ranges are available to you. The meter face in Figure 6-40 has three scales for measuring voltage: one is marked 10, another 50, and the third is marked 250. Remember the scale you read is determined by the position of the range selector knob.

Example 6-7: If the range is set on 250 V, as in Figure 6-41, you read the 0- to 250-V scale.

Reading the AC Voltage Scale Using an Analog Multimeter

When reading the pointer position, be sure to read the line marked AC (Figure 6-40). The spaces on the voltage scales are always equally divided. When the pointer stops between the marks, just read the value of the nearest mark. In Figure 6-42, the pointer is between

Figure 6-39
An example of scales
used on some VOMs.

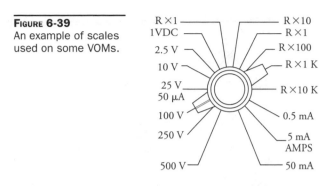

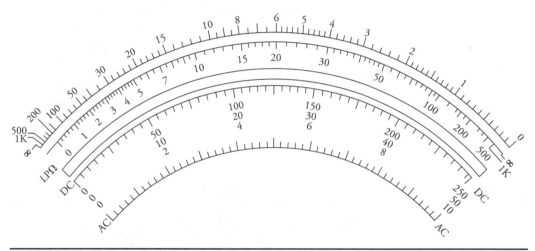

Figure 6-40 The meter face of an analog meter.

115 and 120 V on the 250 scale. Read it as 120 V. With an analog meter, you're not gaining anything by trying to read the voltage exactly.

Measuring Line Voltage Using an Analog Multimeter

Measuring line voltage is the first, and most important, part of checking out air conditioning, heating, or refrigeration equipment that does not operate. Line voltage is the voltage coming from the wall outlet; disconnect box, or circuit breaker panel. The voltage should be approximately the amount of volts AC as stated on the data tag, on the equipment, under "no-load" conditions. No-load means that no air conditioning, heating, or refrigeration equipment is connected or that the product is connected but it is turned off.

To measure line voltage under no-load conditions at 120 V (Figure 6-43):

1. Set the meter to measure AC volts.
2. Set the range selector knob to the range nearest to but higher than 120 V.
3. Insert either test lead into one slot of an empty wall receptacle.

The range is set on
the 250-V scale.

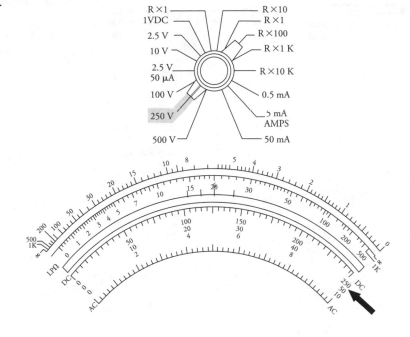

The range is set on the
250-V scale, and the
pointer reads 120 V.

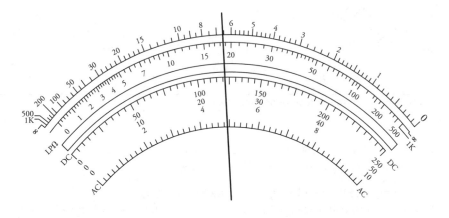

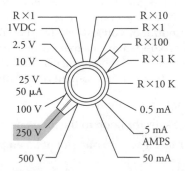

FIGURE 6-43
Measuring line voltage
with no load.

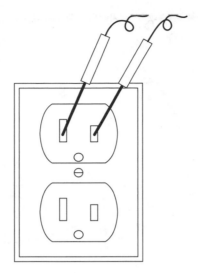

4. Insert the other test lead into the other slot of the same outlet. (Disregard the ground terminal for this test.)

WARNING *Do not touch or handle the test leads by the metal portion of the probe. Hold the probe by the plastic grips that are attached to the test leads to avoid electric shock.*

5. Read the meter. The reading should be between 115 and 120 V.

6. When testing for 240 V, be sure that the range selector knob is set to the nearest range higher than 240 V.

NOTE *Most small BTU-sized air conditioners are rated at 120 V, but will work on voltages ranging from 110 to 125 V. If the voltage drops more than 10%, the air conditioner will not operate, and most likely will damage some electrical components if the air conditioner keeps running.*

To measure line voltage under load at 120 V (Figure 6-44):

1. Be sure that the air conditioner is plugged into one of the receptacles and that the product is turned on.

2. Follow steps 1 to 5 for no-load conditions, inserting the test leads into the empty receptacle next to the one into which the product is plugged.

3. Under load conditions (air conditioner is turned on), your reading will be slightly less than under no-load conditions.

4. When testing for 240 V, be sure that the range selector knob is set to the nearest range higher than 240 V.

NOTE *Products with motors, compressors and heating coils, such as air conditioning, heating, or refrigeration equipment, should also be tested at the moment of start. If the voltage drops more than 10% of the supplied voltage when the equipment is started, there is a problem with the electrical supply.*

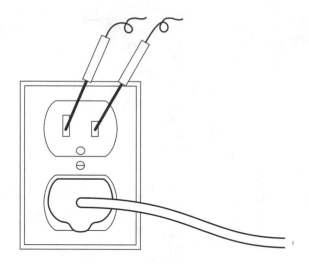

FIGURE 6-44
Measuring line voltage under load.

Not All Digital Multimeters Are the Same

When choosing a digital multimeter (DMM), they are all calibrated to give a root mean squared (rms) indication of the measured signal. The difference between them is how they calculate the final reading showing on the display (LCD) screen. The three commonly used methods are as follows:

- *Peak method.* The DMM reads the peak voltage and then divides the reading by 1.414 (the square root of two) to get the *rms* valve.

- *Averaging method.* The DMM determines the average value of the rectified voltage by using the k value to scale all waveforms the DMM measures. This average value is related to the *rms* value by using the constant k value (k = 1.1).

- *True rms.* DMMs use an integrated circuit within the meter to calculate the *rms* value by squaring the voltage as it samples the voltage reading and then averaging the voltage over a period of time and then taking the square root of the result.

Common Digital Multimeter (DMM) Symbols

Figure 6-45 shows the type of symbols that are often found on DMMs and wiring schematics. These symbols are designed to symbolize the components and reference values. In Figure 6-46, the technician must know where the symbols are located on his meter in order to get a correct reading.

All DMMs come with a battery symbol to indicate when the technician needs to change the batteries in the meter. When the batteries are low or weak, the battery indicator will begin to flash. If the batteries are not replaced, then the meter will not function at all. Some DMMs will show a number 1 on the LCD screen on the left side when the meter is set to measure ohms. This means that the quantity being measured exceeds the maximum value set by the scale; when measuring resistance, this display indicates that current cannot flow due to an open circuit. Simply adjust the range selector knob to a higher range setting to

FIGURE 6-45
Common symbols that
appear on a digital
multimeter LCD
screen.

~	AC Voltage	⎓	Ground
⎓	DC Voltage	⊣⊢	Capacitor
Hz	Hertz	μF	MicroFarad
+	Positive	μ	Micro
—	Negative	m	Milli
Ω	Ohms	M	Mega
⊣⊢	Diode	K	Kilo
•)))	Audible continuity	OL	Overload

FIGURE 6-46
When the DMM is first
turned on all the
symbols and numbers
light up.

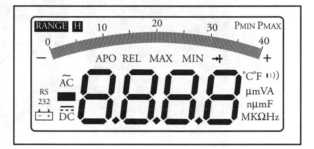

increase the maximum possible reading until you can record a measurement. When the
number 1 appears on the right side of the display, this indicates that you are reading 1 Ω.
This will be a normal measurement.

Reading the Voltage Scales Using a Digital Multimeter

Before you begin testing for voltage, select the correct voltage scale (AC or DC) (Figure 6-47).
Then choose the correct voltage range selection. Always select a range that is higher than the
voltage being tested. If you are unsure as to what type of voltage you should be testing or
how much voltage, read the wiring schematic or service manual of the product you are
servicing.

NOTE *Always select a range that is higher than the voltage you are measuring. If you do not know
the voltage for the product you are servicing, set the range to the highest scale or set the meter to
auto range.*

Measuring Line Voltage Using a Digital Multimeter

When using a digital multimeter, you will measure the voltage as you would if you were
using an analog multimeter (Figures 6-38, 6-43, and 6-44). When reading the LCD screen,
make sure you also read the symbols on the left and on the right side of the numerical
reading to make sure you are reading the correct voltage. If you do not get a proper voltage
reading or the measurement fluctuates, reposition the test probes to a better position on a
clean terminal surface or clean wire connection and retest the voltage.

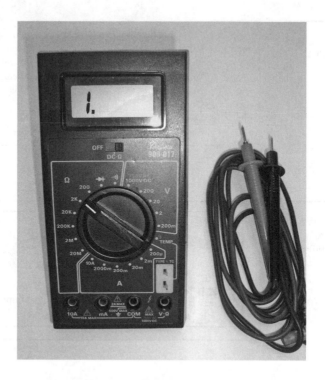

FIGURE 6-47
Choose the functions and ranges correctly before testing live circuits.

Measuring Current using a Digital Clamp-On Multimeter

Most service technicians prefer to use a DMM with a current clamp meter attached as a single meter (Figure 6-48). Some meter manufacturers came out with a current clamp adapter that attaches to a standard DMM (Figure 6-38c). To test for current, press the lever to open the clamp (Figure 6-49), then close the clamp head around a single wire and release the lever. To measure amperage, the product must be operating. Next take a reading, remembering to also read the symbols on the left and right side of the numerical reading.

Measuring Resistance (Ohms) Using an Analog Multimeter

Electrical air conditioning, heating, and refrigeration products need a complete path around which electricity can flow. If there is infinite resistance to the flow of electricity, you have an open circuit or infinite resistance between the two points being measured. When there is a complete path, you have continuity in the circuit. When you test to find out whether there is a break in the path, you say you are making a continuity check.

Continuity checks are made by measuring the amount of resistance there is to the flow of electricity. If there is so much resistance that it is too high to measure (called *infinite*), then you say that the circuit is open (there is no complete path for the electricity to follow). If there is some resistance, it means that there is continuity, but that there is also one (or more) load on the line—a light, a motor, etc.

NOTE   *A load is an electrical component that uses electricity to work (e.g., a light bulb, a motor, or a heater coil).*

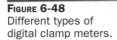

FIGURE 6-48
Different types of
digital clamp meters.

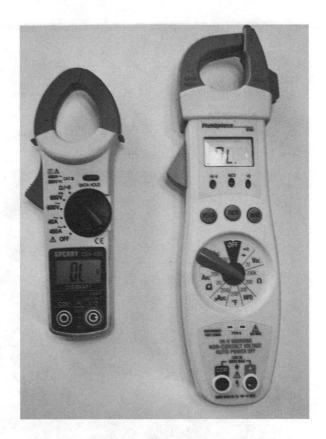

If there is no resistance between the two points, it means that the electricity is flowing directly from one point to the other. If the electricity flowed directly from one point to the other by accident or error, then you say you have a short circuit or a *short*.

Setting Up the Analog Multimeter and Testing for Resistance (Ohms)

Measuring resistance (ohms) is like measuring voltage, except that the measurements are made with the electricity turned off. The steps for measuring ohms are as follows:

1. Attach the leads to the meter. Plug the black lead into the negative outlet and the red lead into the positive outlet.

2. If your meter has a function switch, set the function switch to OHMS.

3. If your meter has a range switch, set the range. The range selector switch will have several ranges of resistance (ohms) measurements (Figure 6-50). The ranges are shown like this:
   - $R \times 1$: The actual resistance shown on the meter face times 1.
   - $R \times 10$: The resistance reading times 10 (add one zero to the reading).
   - $R \times 100$: The resistance reading times 100 (add two zeros to the reading).
   - $R \times 1K$: The resistance reading times 1,000 (add three zeros to the reading).
   - $R \times 10K$: The resistance reading times 10,000 (add four zeros to the reading).

Figure 6-49   To get a correct amperage reading, only place the clamp around a single wire and make sure that the product is operating.

4. Set the range so that it is higher than the resistance you expect. If you don't know what measurement to expect, use the highest setting and adjust downward to a reading of less than 50. The left side of the scale is too crowded for an accurate reading.

5. Zero the meter. You should do this each time you set it. To "zero the meter" means to adjust the pointer so that it reads 0 when the two test leads are touched together. Use the Ohms Adjust knob on the front of the VOM to line up the pointer over the zero on the ohms scale.

6. Attach the test leads to the component you are measuring. Make sure the component or wire is isolated from the remaining circuit to get a true reading.

7. Take the measurement. In Figure 6-50, the range selector switch is on $R \times 100$, and the measurement is 400 Ω.

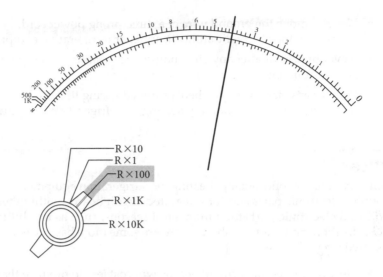

FIGURE 6-50
With the range set on
R × 100, the meter
reads 400 Ω.

Measuring Resistance (Ohms) with a Digital Multimeter

Before testing for resistance, turn off the electricity to the air conditioning, heating, or refrigeration equipment. If you test a circuit or load in the circuit with the electricity on, you could damage or destroy your meter. Testing for resistance with a DMM is the same as testing with an analog VOM. It is important to have good contact between the meter test probes and the component being tested. Take the measurement, and read the numerical value and symbols on the left and right side of the reading.

Electrical Safety Precautions

Know where and how to turn off the electricity to your air conditioning, heating, or refrigeration equipment, for example, plugs, fuses, circuit breakers, disconnect boxes, or cartridge fuses. Know their location in the home or business and label them. When replacing parts or reassembling the air conditioning, heating, or refrigeration equipment you should always install the wires on their proper terminals according to the wiring diagram. Then check to be sure that the wires are not crossing any sharp areas, are pinched in some way, or are between panels or moving parts that might cause an electrical problem.

These additional safety tips can help you and your family:

- Always use a separate, grounded electrical circuit for each air conditioning, heating, or refrigeration equipment.

- If you have to use an extension cord for a room air conditioner, make sure to use one that is properly rated for the size air conditioner installed.

- Be sure that the electricity is off before working on air conditioning, heating, or refrigeration equipment.

- Never remove the ground wire of a three-prong power cord, or any other ground wires from the air conditioning, heating, or refrigeration equipment.
- Never bypass or alter any air conditioning, heating, or refrigeration equipment switch, component, or feature.
- Replace any damaged, pinched, or frayed wiring that might be discovered when repairing the air conditioning, heating, or refrigeration equipment.

Electronics

When servicing air conditioning, heating, or refrigeration equipment, you will also encounter electronic parts and/or integrated circuit boards within them. Manufacturers introduced electronic parts for a number of reasons, such as reliability, miniaturization, standardization, and maintainability. Here are some more basic reasons for electronic integrated circuit boards:

- Electronic integrated circuit boards will cost less to produce than its mechanical counterpart.
- You can program and/or monitor the electronic integrated circuit board to save the customer money during operation.
- The customer can choose more program options that come with the component.
- To make the product more efficient, the electronic control can use the feedback circuits.
- Electronic integrated circuit board displays are easier to read and set than the old fashion mechanical controls.
- During power outages, some electronic integrated circuit boards have a battery back up to keep track of what was programmed into the board, including the date and time.
- Electronic integrated circuit boards have built-in self-diagnostics and error codes.

When troubleshooting these electronic parts or integrated circuit boards, the technician will troubleshoot and determine that an integrated circuit board has failed and will change the entire circuit board rather than repair the electronic components on the board. Although this chapter cannot cover all there is to know about electronics, it will provide the basics. With the influx of electronic components into air conditioning, heating, and refrigeration equipment, it is imperative that service technicians be able to troubleshoot them.

In order to troubleshoot electronic circuits the technician will need the following:

- The technical data sheet or wiring diagram.
- Understanding of the air conditioning, heating, or refrigeration equipment operation.
- Obtain from the customer how the controls were set and usage of the product prior to your arrival.
- Knowledge of how to access and use the diagnostic service mode. Each product has a different way to access the service mode.
- Understanding of how to read wiring diagrams and other circuit charts.

In order to properly diagnose a problem in an air conditioning, heating, or refrigeration equipment, you will have to determine if the problem is either, both or all of the following:

- Electrical
- Mechanical
- Customer related

Before you can replace an electronic component, you must rule out everything else.

Integrated Circuits and Circuit Boards

An integrated circuit (IC), shown in Figure 6-51a and 6-51b, is an electric circuit miniaturized and consisting of transistors, diodes, resistors, capacitors, and all the connecting wiring—all of it manufactured on a single semiconductor chip. Integrated circuits are found in products such as computers, televisions, cell phones, and major appliances and air conditioners. IC chips are not repairable.

A printed circuit board (PCB) is a layout of conductive pathways (an electric circuit) secured to a board. Electronic components are soldered on to the board for a particular function within an appliance or air conditioner. Only a small percentage of printed circuit boards are repairable. The remaining percentage of PCBs is replaced with new boards.

Circuit Board

When diagnosing the circuit board, the wiring schematic for the appliance will be helpful in diagnosing, understanding wire color codes, and reading the correct voltages. For example, in Figure 6-52, the main PCB controls the on/off functions and the temperature for the air conditioner. To determine if the main PCB is defective, you would check for the correct supply voltage coming into the PCB. In this case, the voltage should be 120 V AC. In addition, you can also check the voltage at the primary winding of the transformer mounted on the PCB for 120 V AC. Next, test the secondary side of the transformer for output voltage. There is a line fuse on this PCB. Turn off the electricity and check the fuse for continuity. If all checks out, test the relays on the PCB for voltage to the relay coils, or test if the switch contacts on the relays are opening and closing.

A good rule to remember when testing any PCB is that there must be voltage supplied to the board and there must be voltage leaving the board to turn a function on.

Transformer

A transformer is an electrical device that can increase (step up) or decrease (step down) the voltage and current. It works on the principle of transferring electrical energy from one circuit to another by electromagnetic induction (Figure 6-53). The primary side of the transformer is the high-voltage side, with the voltage ranging from 120 V AC to 240 V AC. On the secondary, or low-voltage, side, the voltage will range from 5 V AC to 24 V AC, depending on the amount of voltage and current needed to operate the circuit boards. Some circuit boards require DC voltage to operate, depending on the manufacturer's requirements for the product.

(a)

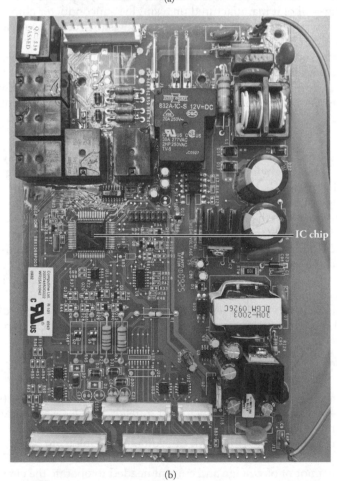

IC chip

(b)

FIGURE 6-51   (a) An integrated circuit (IC) chip and (b) an integrated circuit (IC) chip on PC board.

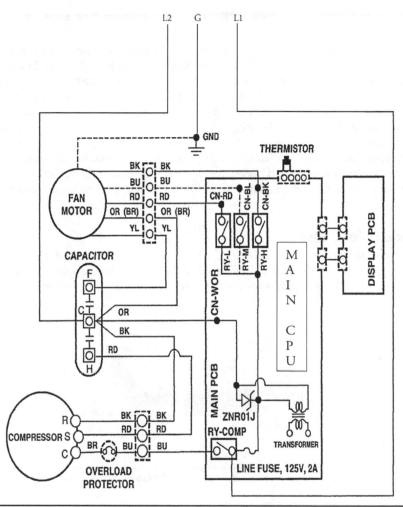

FIGURE 6-52 A sample RAC wiring schematic.

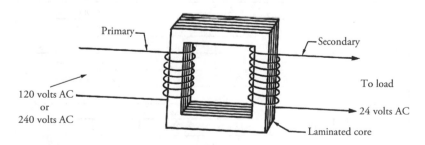

FIGURE 6-53 Most appliances and air conditioners use a step-down transformer to supply a low voltage to electronic PCBs.

Resistors

A resistor, when installed into an electrical circuit, will add resistance, which will produce a specific voltage drop, or a reduction in current. Resistors can be either fixed or variable. The most common problem found with resistors when testing them with a multimeter is that they read an open (infinite ohms). To determine the resistance value in ohms of a resistor, you have to isolate it from the circuit and then use your resistance scale (ohms) to read its value. On larger resistors, the resistance value is stamped on the body of the resistor.

A visual inspection of the resistor should also be performed to rule out any physical flaws in the resistor and the surrounding components. If you touch a resistor, and it begins to flake apart, most likely it is bad and it needs to be replaced. If the technician smells a burning smell coming from the resistor, most likely this component is bad and it must be replaced.

Resistance Color Bands

Because of their size, most axial resistors (Figure 6-54a) are color-coded, usually with four bands to indicate their resistance value. Reading the resistor bands from left to right, the first band closest to the end indicates the first numerical digit of resistance. The second

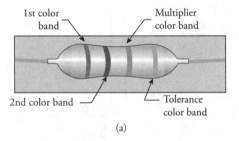

(a)

Color	1st Band	2nd Band	Multiplier
Black	0	0	× 1
Brown	1	1	× 10
Red	2	2	× 100
Orange	3	3	× 1000
Yellow	4	4	× 10,000
Green	5	5	× 100,000
Blue	6	6	× 1,000,000
Violet	7	7	× 10,000,000
Gray	8	8	× 100,000,000
White	9	9	× 1,000,000,000

4th Band = Tolerance							
Gold	5%	Silver	10%	No Color	20%

(b)

Figure 6-54 (a) A common resistor and (b) resistor color-code chart.

band indicates the next numerical digit of resistance, and the third band indicates the numerical multiplier in zeros. The fourth band on a resistor is the resistance tolerance measured in percentage (Figure 6-54b). For example, let's say the color bands are brown, red, orange, and silver:

Brown = 1

Red = 2

Orange = 1000

Silver = + or −10%

The resistor is rated at 12,000 Ω with a ±10% tolerance. The tolerance is the safe operating percentage difference of the color-coded resistance value. For instance, a 12,000-Ω resistor with a 10% tolerance will operate safely at 10% below or 10% above the 12,000-Ω valve.

Thermistor

Thermistors are constructed from a semiconductor material with a specified temperature range. This electronic device exhibits a change in resistance with a change in temperature. There are two types of thermistors: negative temperature coefficient (NTC) and positive temperature coefficient (PTC), with the most common being NTC. The resistance of the NTC thermistor decreases with an increase of temperature, whereas the resistance of the PTC thermistor increases with an increase of temperature. There are many applications for thermistors, including their use as temperature sensors, resettable fuses, current limiters, and power indicators. They are also used in applications where interchangeability without recalibration is required.

One type of test performed on thermistors to check if they are good or bad is the resistance-versus-temperature test. To perform this test, the technician will have to locate the technical data sheet that comes with the product. The data sheet will provide the resistance (ohms) valve at the test temperature listed. Another way to test the thermistor is by testing it with an ohmmeter for an open or a short.

Diodes

A diode is an electrical device that allows current to flow in one direction only. If you connect a positive voltage to the anode (positive side of the diode) and a negative voltage to the cathode side (negative side of the diode), the diode becomes forward-biased (Figure 6-55a). When the diode is forward-biased, current will flow. When a positive voltage is connected to the cathode (negative side of the diode) and a negative voltage is connected to the anode (positive side of the diode), the diode becomes reverse-biased (Figure 6-55b). Then the diode will not allow current to flow. A diode may be thought of as a switch "closed" when forward-biased and "open" when reverse-biased. There are many types of uses for diodes besides rectification. These include capacitance that varies with the amount of voltage applied to the diode and photoelectric effects. The wiring diagram symbol for a diode is shown in Figure 6-56.

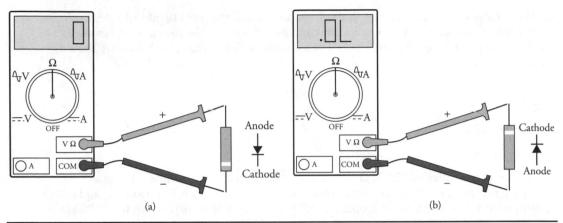

FIGURE 6-55   (*a*) When testing the diode with an ohmmeter, the forward bias indicates zero ohms of resistance; (*b*) when testing the diode with an ohmmeter, the reversed bias indicates infinite resistance.

When a diode is connected in a circuit, there must be a means to identify the cathode and the anode (Figure 6-57). Diodes are manufactured in different case styles. On large diodes, there is a diode symbol stamped on the side to indicate to the technician the cathode and anode. On smaller diodes, there is a band around one end to identify the cathode.

Testing a Diode

A diode can be tested with an ohmmeter. To test a diode, you must first disconnect one side of the diode to isolate it from the remainder of the circuit. Set your ohmmeter to the ohms scale, and connect the leads to the diode. The meter may or may not read continuity. Then reverse the leads on the diode and check for continuity. You should read continuity in one direction only. If continuity is not indicated in either direction, the diode is open (Figure 6-58*a*). If continuity is indicated in both directions, the diode is shorted (Figure 6-58*b*).

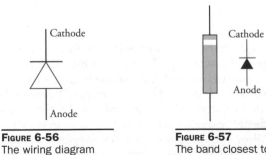

FIGURE 6-56
The wiring diagram
symbol for a diode.

FIGURE 6-57
The band closest to the
end of the diode is
the cathode (–) and the
other end is known as
the anode (+).

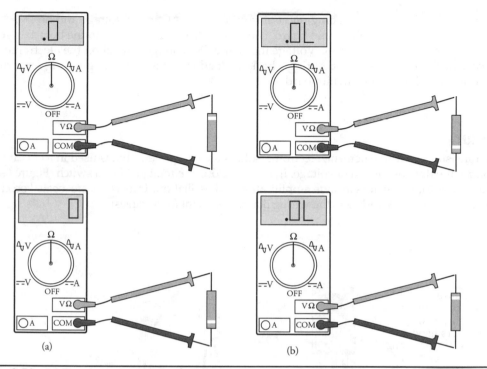

FIGURE 6-58   (a) The ohmmeter indicates no continuity in either direction. The diode is open; (b) the ohmmeter reads continuity in both directions, indicating a shorted diode.

Bridge Rectifier

The bridge rectifier (Figure 6-59), consists of four diodes connected together in a bridge configuration on the circuit board. On electronic control boards, a bridge rectifier is used to convert alternating current into direct current, for the low voltage circuitry.

The technician cannot test the bridge rectifier directly because it is wired directly into the circuitry on the integrated circuit board. You can only test the AC input voltage and DC output voltages. If you're reading on your voltmeter, AC input voltage to the electronic board and no DC output voltage from the board, the board is bad and will need

FIGURE 6-59
A bridge rectifier circuit converting AC to DC.

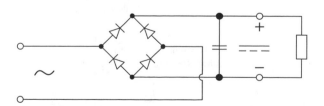

to be replaced. If you're reading on your voltmeter, no AC input voltage and no DC output voltage on the electronic board, check the AC voltage supply to the board. If your reading on your voltmeter, AC voltage input and DC voltage output on the electronic board, and the load is not operating, check the load and the connecting wiring form the load to the relay on the circuit board.

Transistor

A transistor is a three-element, electronic, solid-state component that is used in a circuit to control the flow of current or voltage. It opens or closes a circuit just like a switch (Figure 6-60a). Other uses for transistors include amplification and oscillation. Transistors are located on circuit boards. Figure 6-60b indicates the wiring diagram symbol for a transistor.

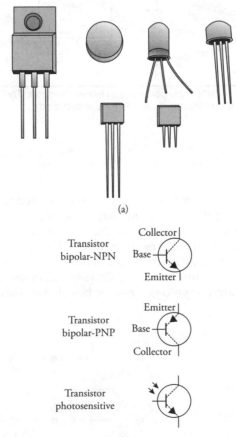

FIGURE 6-60
(a) A sample of the various types of transistors used in circuits; (b) the wiring symbol for an NPN, PNP, and photosensitive transistor.

(a)

Transistor
bipolar-NPN

Collector
Base
Emitter

Transistor
bipolar-PNP

Emitter
Base
Collector

Transistor
photosensitive

(b)

LED Multiple-Segment Displays

LED segment displays are available in 7, 9, 14, and 16 segment displays (Figure 6-61). Each segment will light up to produce an alphanumeric character. Applications today include displays fitted for refrigerators and air conditioners. Light-emitting diodes (LEDs) are a semiconductor light source, and they are available in a variety of shapes, sizes, and colors. LEDs are used as a lamp indicator. The LED was first introduced for electronics in 1962.

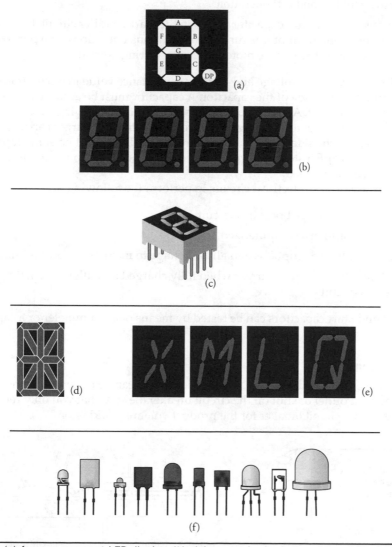

FIGURE 6-61   (a) A seven-segment LED display; (b) alphanumeric characters; (c) an LED; (d) and (e) sixteen-segment LED displays used on major appliances and A/Cs; and (f) light-emitting diodes (LEDs) are a semiconductor light source.

Capacitors

A capacitor is a device that stores electricity to provide an electrical boost for motor starting (Figure 6-62). Most high-torque motors need a capacitor connected in series with the start winding circuit to produce the desired rotation under a heavy starting load.

There are two types of capacitors:

- *Start capacitor*. This type of capacitor is usually connected to the circuit between the start relay and the start winding terminal of the motor. Start capacitors are used for intermittent (on and off) operation.

- *Run capacitor*. The run capacitor is also in the start winding circuit, but it stays in operation while the motor is running (continuous operation). The purpose of the run capacitor is to improve motor efficiency during operation.

Capacitors are rated by voltage and by their capacitance value in microfarads (µF). This rating is stamped on the side of the capacitor. A capacitor must be accurately sized to the motor and the motor load. Always replace a capacitor with one having the same voltage rating and the same (or up to 10% greater) microfarad rating. On larger capacitors, the rating is stamped on the side. Also, watch out for the decimal point on some capacitors. The rating might read .50 µF instead of 50 µF. Small capacitors in electronic circuits are rated by numbering or are color-coded.

Capacitors are used in electrical circuits to perform the following:

- An electrical voltage boost in a circuit
- Control timing in a computerized circuit
- Reduce voltage disruptions and allow voltage to maintain a constant flow
- Block the flow of direct current when fully charged and allow alternating current to pass in a circuit

Both run and start capacitors can be tested by means of an ohmmeter or a capacitor tester.

Testing a Capacitor

Before testing a capacitor, disconnect the electricity. This can be done by pulling the plug from the electrical outlet or shut off the circuit breaker. Be sure that you only remove the plug or shut off the circuit breaker for the product you are working on.

FIGURE 6-62
(a) The capacitor is rated by voltage and by capacitance (in microfarads); (b) this built-in disconnect device is also known as a fail-safe.

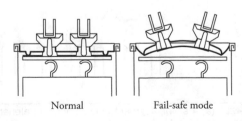

Normal   Fail-safe mode

(a)   (b)

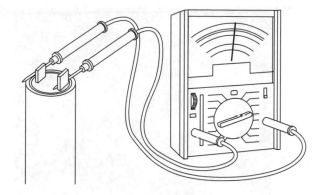

FIGURE 6-63
Placing ohmmeter test
leads on the capacitor
terminals.

Some air conditioning or refrigeration product models have the capacitor mounted on the motor, and some are mounted to the cabinet interior in the rear of the machine. Access might be achieved through the front or rear panel, depending on which model you are working on. Do not touch the capacitor until it's discharged.

WARNING *A capacitor will hold a charge indefinitely, even when it is not currently in use. A charged capacitor is extremely dangerous. Discharge all capacitors immediately any time that work is being conducted in their vicinity. Redischarge after repowering the equipment if further work must be done.*

Before removing the wires from the capacitor, use a screwdriver with an insulated handle to discharge a larger capacitor by shorting it across both terminals. Remove the wire leads one at a time with needle nose pliers. Set the ohmmeter on the highest scale, and place one probe on one terminal and the other probe on the other terminal (Figure 6-63). Observe the meter action. While the capacitor is charging, the ohmmeter will read nearly 0 Ω for a short period of time. Then the ohmmeter reading will slowly begin to return toward infinity. If the ohmmeter reading deflects to zero and does not return to infinity, the capacitor is shorted and should be replaced. If the ohmmeter reading remains at infinity and does not dip toward zero, the capacitor is open and should be replaced.

Another way to test a capacitor is to use a capacitor tester. When you use this type of meter, you will be able to check if you have a weak capacitor. If the reading in microfarad is below the rating microfarad on the capacitor, replace the capacitor.

Heating Elements

Most heating elements are made with a nickel-chromium wire, having both tensile strength and high resistance to current flow. The resistance and voltage can be measured with a multimeter to verify if the element is functioning properly. Heating elements are available in many sizes and shapes (Figure 6-64). They are used for environmental heating, defrost for refrigerators and freezers.

Heating elements are not repairable, and they should be replaced with a duplicate of the original.

(a) Nickel-chromium wire, heater

(b) Calrod heater

FIGURE 6-64 Heating elements.

Electrostatic Discharge

The electronic parts in an air conditioning, heating, or refrigeration equipment are sensitive to electrostatic discharge (ESD). An example of electrostatic discharge occurs when you rub your feet on a carpet and then touch a metal doorknob. The little spark that you generated by touching the metal doorknob is ESD. This type of electrostatic discharge will damage the electronics in the air conditioning, heating, or refrigeration equipment at the microscopic level. When servicing an air conditioner or installing a new electronic part, ESD may cause failure in the future due to stress from these occurrences.

To prevent ESD from damaging expensive electronic components, follow these steps:

- Turn off the electricity to the air conditioning, heating, or refrigeration equipment before servicing any electronic component.

- Before servicing the electronics in an air conditioning, heating, or refrigeration equipment, discharge the static electricity from your body by touching your finger repeatedly to an unpainted surface on the product. Another way to discharge the static electricity from your body is to touch your finger repeatedly to the green ground connection on that product.

- The safest way to prevent ESD is to wear an anti-static wrist strap.

- When replacing a defective electronic part with a new one, touch the anti-static package that the part comes in to the unpainted surface of the equipment or to the green ground connection of the air conditioning, heating, or refrigeration equipment.

- Always avoid touching the electronic parts or metal contacts on an integrated board.

- Always handle integrated boards by the edges.

Wires

The wiring, which connects the different components in an air conditioning, heating, or refrigeration products is the highway that allows current to flow from point A to point B. Copper and aluminum are the most common types of wires that are used in air conditioning, heating, and refrigeration equipment. They are available as solid or stranded. Wires are enclosed in an insulating sleeve, which might be rubber, cotton, or one of the many plastics. Wires are joined together or to the components by:

- Solderless wire connectors

- Solderless wire terminal connectors

- Solderless multiple-pin plug connectors

- Soldering

Never join copper and aluminum wires together, because the two dissimilar metals will corrode and interrupt the flow of current. The standard wire-gauge sizes for copper wire are listed in Table 6-6. As the gauge size increases from 1 to 20, the diameter decreases and the amperage capacity (ampacity) will decrease also (see Table 6-7).

Gauge No.	Diameter, Mil	Circular-Mil Area	Ohms Per 1000 ft. of Copper Wire at 25°C*
1	289.3	83,690	0.1264
2	257.6	66,370	0.1593
3	229.4	52,640	0.2009
4	204.3	41,740	0.2533
5	181.9	33,100	0.3195
6	162.0	26,250	0.4028
7	144.3	20,820	0.5080
8	128.5	16,510	0.6405
9	114.4	13,090	0.8077
10	101.9	10,380	1.018
11	90.74	8234	1.284
12	80.81	6530	1.619
13	71.96	5178	2.042
14	64.08	4107	2.575
15	57.07	3257	3.247
16	50.82	2583	4.094
17	45.26	2048	5.163
18	40.30	1624	6.510
19	35.89	1288	8.210
20	31.96	1022	10.35

TABLE 6-6   Standard Wire-Gauge Sizes for Copper Wire

Size (AWG)	Ampacity
18	6
16	8
14	17
12	23
10	28
8	40

TABLE 6-7   Wire Size and Ampacity

How to Strip, Splice, Solder, Install Solderless and Terminal Connectors, and How to Use Wire Nuts on Wires

To strip the insulation off the wire, there are certain steps to follow. First you must have a good wire stripper (Figure 6-65). Now, place the wire in the proper sized slot in the wire stripper and work the stripper back and forth until a cut is made in the entire insulation. Do not damage any of the strands in a stranded wire or put a knick in a solid wire; this will cause a weakness in the wire that may cause a break in the circuit in the not too distant future. To remove the insulation, hold the wire tight with one hand and use the other hand to gently move the insulation back and forth until the cut breaks clean and the unwanted insulation can be pulled off the wire (Figure 6-66). Next, taper the insulation with a knife to increase the wire's flexibility because a straight cut in the insulation will create a force that can cause a wire to break prematurely (Figure 6-67). Figures 6-68 and 6-69 illustrate the different methods of splicing single and stranded wires together.

To connect wires to screw terminals, the wire being attached at a screw terminal should be attached to the screw terminal so that the loop lies in the direction the screw turns

Figure 6-65
Always use a wire stripper to remove the insulation from the wire.

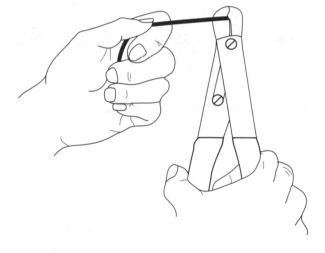

Figure 6-66
Move insulation back and forth, to remove insulation from wire.

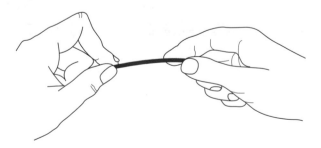

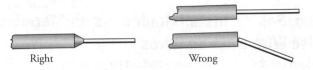

Right   Wrong

FIGURE 6-67   Taper the insulation on the wire with a knife to increase the wires flexibility.

Types of Splices

A.   Simple Splice

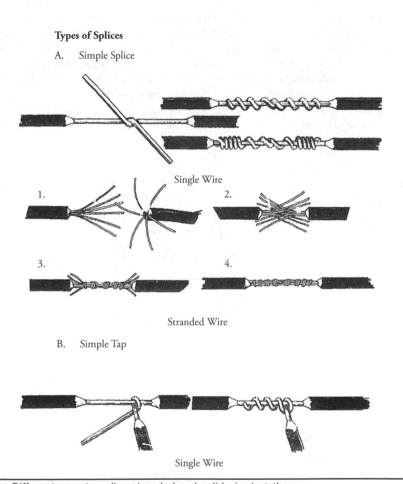

Single Wire

Stranded Wire

B.   Simple Tap

Single Wire

FIGURE 6-68   Different ways to splice stranded and solid wire together.

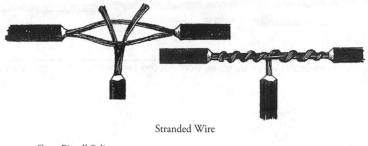

Stranded Wire

C.   Pigtall Splice

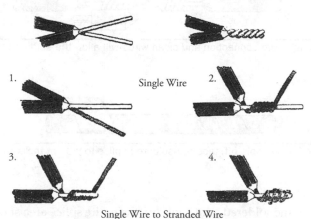

1.   Single Wire   2.

3.   4.

Single Wire to Stranded Wire

D.   Hook Splice

FIGURE 6-69   Different ways to correctly splice wire together.

Right   Wrong

FIGURE 6-70   Connecting a wire correctly to a screw terminal.

(Figure 6-70). The wire should loop the screw a little less than one full turn, but excessive loops around the screw terminal is not recommended, it could cause wire damage.

Here is a good guideline that you should practice for soldering wire splices (Figures 6-71 to 6-73). First, the wire will be soldered together, should be bright and clean at the point of connection. The connection point should be tight together so that the solder can flow between the joint to solidify without any wire movement. When soldering wires together, the wires should be coated with an electric soldering paste of flux, and soldered so that the solder melts and flows into every crevice of

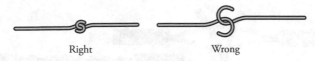

Right   Wrong

FIGURE 6-71   Preparing to solder wires together. Connect the two ends of the wires as shown.

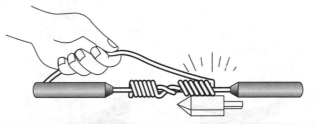

FIGURE 6-72   Having a tight wire connection and clean wires will allow the solder to flow through the joint.

FIGURE 6-73   When the solder cools off, tape the soldered splice to protect the wire from shorting out.

the spliced joint. After the soldered joint cools off, the entire splice area should be covered with a waterproof plastic tape or heat shrink covering to protect the joint from shorting out against the cabinet of the product.

When you have to attach a solderless connector or wiring terminal to a wire you should follow these guidelines. Solderless connectors should be used according to their color codes, and a connector for a smaller gauge wire should never be used on a heavier gauge wire. The wire connector might burn off the wire and break the circuit. A screw-on wire connector (also known as a wire nut) work well on pigtail splices (Figure 6-74).

When installing a crimp-on connector (Figure 6-75), there should not be a gap between the insulation and the terminal connector, and if there is a gap, a plastic sleeve should be

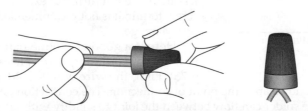

FIGURE 6-74   Wire connector also known as a wire nut. Twist wires in a clockwise direction; screw the wire nut cap in a clockwise direction onto the connected wires as shown.

FIGURE 6-75 When attaching terminal connectors, try not to have any gaps from the end of the insulation to the end of the wire connector.

added to cover the bare wire, or reinstall a new connector if needed. Always prevent wires from shorting out, breaking the circuit, causing you a callback on the call.

Circuit Protection Devices

Circuit protection devices are important for air conditioning, heating, and refrigeration equipment. These devices will protect the electrical circuits and components from damage from too much current flow. Each fuse or circuit breaker (Figures 6-76 to 6-78) must be rated for voltage and current (Table 6-8). Never replace a fuse or circuit breaker with one that is not correctly rated for the product. Fuses and circuit breakers must be able to do the following:

- Sense a short in the circuit
- Sense an overloaded circuit (too much current)
- Will not open a circuit under normal operating current conditions
- Opens a circuit before electrical damage to components or product
- Should not exceed the current carrying capacity of the circuit

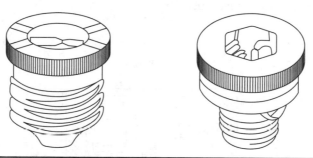

FIGURE 6-76 Two types of screw-in fuses.

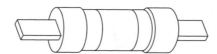

FIGURE 6-77 A cartridge fuse.

FIGURE 6-78
A circuit breaker.

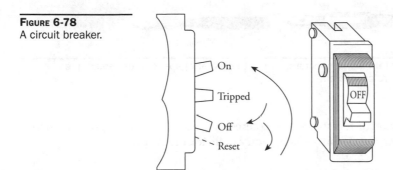

L	Time delay
RK1	Time delay; fast acting
RK5	Time delay
T	Fast acting
J	Time delay; fast acting
CC	Time delay; fast acting
CD	Time delay
G	Time delay
K5	Fast acting
H	Renewable fuse; fast acting

TABLE 6-8 Cartridge Fuses Are Classified as One-Time Use, Renewable, Dual-Element, Current-Limiting, or High Interrupting Capacity, by Underwriters Laboratories (UL)

FIGURE 6-79
A thermal overload.

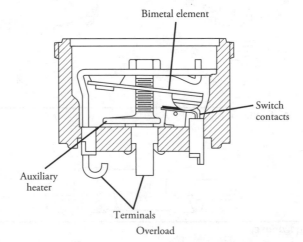

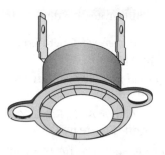

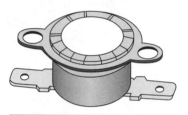

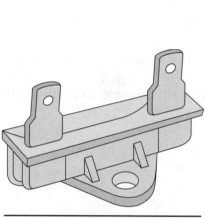

FIGURE 6-80   A thermal fuse.

FIGURE 6-81   A bimetal thermostat.

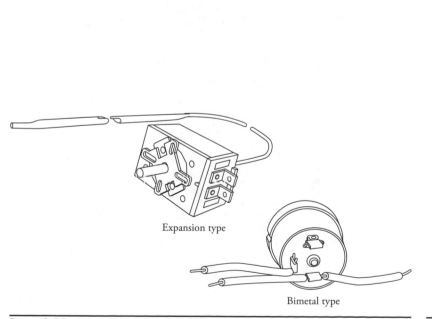

Expansion type

Bimetal type

FIGURE 6-82   Bimetal and expansion thermostats.

FIGURE 6-83   A fusible link also known as a fuselink.

Other circuit protection devices include:

- *Thermal overload.* Operated by heat usually resettable, can also be one time use (Figure 6-79).
- *Thermal fuse.* Incased in an insulated case, one time use, turns circuit off (Figure 6-80).
- *Bimetal thermostats.* Resettable switches, rated by temperature change, turns circuit on or off (Figure 6-81).
- *Bimetal switches.* Turns a component on or off by design temperature rating (Figure 6-82).
- *Fusible links.* Has a one time thermal limit, turns the circuit off (Figure 6-83).

CHAPTER 7

Gas

As a technician or novice servicer, this chapter on gas is very important to the safety and technical knowledge that is needed to repair gas furnaces and gas boilers. In addition, consumers should be knowledgeable of safety procedures and basic characteristics of natural and liquefied petroleum gas (LP or LPG). Any person who cannot use basic tools or follow written instructions should not attempt to install, maintain, or repair gas appliances.

If you do not fully understand the procedures in this chapter, or if you doubt your ability to complete the task on your gas product, please call your service manager.

Currently, gas is used in millions of homes for heating, cooling, cooking, drying laundry, and water heating. Understanding gas theory, gas conversion, ignition systems, and the different types of gases used is something that every technician needs in his arsenal to service gas products. In addition, the technician needs to understand two other important elements when servicing gas products: combustion and ventilation. Some products use only gas as the main fuel, while others use gas and electricity. Although this chapter cannot cover all there is to know about gas, it will provide the basics.

Gas Safety

The following are a few safety tips to help you in handling gas appliances in your home or office:

- Always follow the manufacturer's installation instructions and the use and care manual for the gas appliance.
- Always keep combustible products away from gas appliances.
- Keep your gas appliance clean.
- Teach your children not to play near or with gas appliances.
- Always have a fire extinguisher nearby just in case of mishaps that might lead to a fire.
- Have a smoke detector and a carbon monoxide detector installed in the home or office, and check the batteries yearly.
- Make sure that gas appliances have proper venting according to the manufacturers' recommendations.
- Have all gas leaks repaired no matter how small they are.

Types of Gas

The two most common gases that are used in homes today are natural gas and liquefied petroleum gas (LP or LPG). Gas is a form of chemical energy, and when it is converted by combustion, it becomes heat energy. This type of heat energy is used for cooking, drying, heating, cooling, and lighting.

Natural Gas

Natural gas is a naturally occurring product made up of hydrocarbon and non-hydrocarbon gases. The main ingredient found in natural gas is methane (70 to 90%), with the remainder of the ingredients being nitrogen, ethane, butane, carbon dioxide, oxygen, hydrogen sulphide, and propane. These gases are located beneath the earth and can be removed through constructed wells. Another method for producing and harvesting natural gas is through landfills. The methane gas that is produced by the decomposition of materials can be harvested and added to the natural gas supply.

The heating value of natural gas is between 900 and 1,200 BTU/ft^3. The air we breathe has a specific gravity of 1.00, and natural gas is lighter than air, with a specific gravity varying from 0.58 to 0.79. Natural gases are odorless and colorless—gas companies add an odor agent to warn for leaks.

The pressure of natural gas that is supplied to a residence will vary, between a 5- and 9-in water column. A gas pressure regulator that is connected to the appliance will further reduce the pressure. For example, on most gas furnaces, the pressure will be reduced to a 3.5-in water column. To determine the correct pressure rating, the technician must refer to the manufacturer's specifications or the installation instructions for that product.

Liquefied Petroleum Gas

Liquefied petroleum gas is obtained from natural gas sources or as a by-product of refining oil. LP gas for domestic use is usually propane, butane, or a mixture of the two. This type of gas is compressed and stored in storage tanks under pressure in a liquid state at approximately 250 lb/in^2. The pressure in an LP tank will vary according to the surrounding temperatures and altitude. LP gas tanks can be transported to areas that are not supplied by natural gas supply lines.

The heating value of propane gas is 2,500 BTU/ft^3, with a specific gravity of 1.53. The heating value of butane is much higher—about 3,200 BTU/ft^3, with a specific gravity of 2.0. Liquefied petroleum gas is heavier than air and will accumulate in low-lying areas on the floor, in an enclosure, or in pockets beneath the ground, creating a hazard if it encounters an open flame. As mentioned, these gases are odorless and colorless, and gas companies add an odor agent to warn for leaks.

LP or LPG pressure for residential appliances, as established by the gas industry, will be between a 9 and 11-in water column. On most LP gas furnaces, the pressure will be reduced between 9 and 10-in water column. To determine the correct pressure rating, the technician must refer to the manufacturer's specifications or the installation instructions for that product.

Combustion

Combustion is a rapid chemical reaction (burning) of LP or natural gas and air to produce heat energy and light. To sustain combustion in a gas appliance, an ignition source—such as that produced by a flame or by electrical means—is used to ignite the

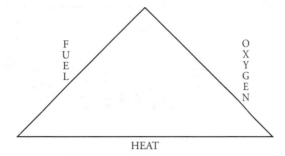

Figure 7-1
The combustion triangle.

F
U
E
L

O
X
Y
G
E
N

HEAT

gas vapors. Three elements are needed to produce a flame when burning gas vapors: fuel, heat, and oxygen. If any one element is missing, flame or burning of gas vapors will not exist (Figure 7-1).

It takes 1 ft³ of gas mixed with 10 ft³ of air to have complete combustion of natural gas. The process produces approximately 11 ft³ of combustion product, consisting of approximately 2 ft³ of water vapor, 1 ft³ of carbon dioxide, 8 ft³ of nitrogen, and the excess air from the gas appliance. These combustion products must be properly vented or discharged safely from the gas appliance.

It takes 1 ft³ of gas mixed with 24 ft³ of air for complete combustion to happen with propane gas. The process produces approximately 25 ft³ of combustion product. With butane gas, it takes 1 ft³ of gas mixed with 31 ft³ of air and produces 32 ft³ of combustion product (Figure 7-2). With the proper mixture of air, gas, and flame, complete combustion

Properties of		Natural gas	Propane	Butane
Chemical formula		CH_4	C_3H_8	C_4H_{10}
Boiling point of liquid at atmospheric pressure	°F	−258.7	−44	32
Specific gravity of vapor (air = 1)		0.6	1.53	2.00
Specific gravity of liquid (water = 1)		0.6	0.51	0.58
Calorific value @ 60°F	BTU/cubic foot	1012	2516	3280
	BTU/gallon		91,690	102,032
	BTU/lb		21,591	21,221
Latent heat of vaporization	BTU/gallon	712	785.0	808.0
Liquid weight	lbs/gallon	2.5	4.24	4.81
Vapor volume from 1 gallon of liquid at 60°F	Cubic foot		36.39	31.26
Vapor volume from 1 lb of liquid at 60°F	Cubic foot		8.547	6.506
Combustible limits	% of gas in air	5–15	2.4–9.6	1.9–8.6
Amount of air required to burn 1 cubic foot of gas	Cubic foot	9.53	23.86	31.02
Ignition temperature in air	°F	1200	920–1020	900–1000
Maximum flame temperature in air	°F	3568	3595	3615
Octane number		100	Over 100	92
All data are approximate.				

Figure 7-2   Properties of utility gases.

FIGURE 7-3
Composition of air.
The air we breathe is
made up of 99.99% of
nitrogen, oxygen,
carbon dioxide, and
argon.

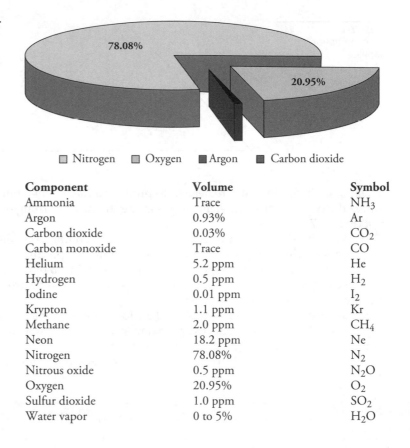

Component	Volume	Symbol
Ammonia	Trace	NH_3
Argon	0.93%	Ar
Carbon dioxide	0.03%	CO_2
Carbon monoxide	Trace	CO
Helium	5.2 ppm	He
Hydrogen	0.5 ppm	H_2
Iodine	0.01 ppm	I_2
Krypton	1.1 ppm	Kr
Methane	2.0 ppm	CH_4
Neon	18.2 ppm	Ne
Nitrogen	78.08%	N_2
Nitrous oxide	0.5 ppm	N_2O
Oxygen	20.95%	O_2
Sulfur dioxide	1.0 ppm	SO_2
Water vapor	0 to 5%	H_2O

will take place. The combustion product that is produced from complete combustion will be carbon dioxide and water vapor.

Inadequate venting of a gas appliance will restrict the flow of air into the gas appliance. This lack of proper ventilation will reduce the amount of oxygen within the air for complete combustion to occur. The air that we breathe is a mixture of gases containing nitrogen, oxygen, argon, carbon dioxide, water vapor, and other trace gases (Figure 7-3). Incomplete combustion will cause the reduction of oxygen levels in the air supply within a room. Proper ventilation—adequate fresh air into the room—is important and cannot be stressed enough for the proper operation of a gas appliance, as well as for the safety of human life within the home.

Carbon Monoxide

Carbon monoxide (CO) is a toxic gas that can cause death if inhaled in copious amounts. It is odorless, colorless, and has no taste. CO cannot be detected without a proper meter. The specific gravity of carbon monoxide is 0.980, which is slightly lighter than air. Air has a specific gravity of 1. CO gas will disperse rapidly and evenly in a home and it can be carried readily with convection air currents within a closed home. The human body cannot detect carbon monoxide with its senses. When carbon monoxide is inhaled, it is absorbed into the bloodstream and stays there longer, preventing oxygenated blood from performing its job in the body. Two factors affect the amount of carbon monoxide absorbed into the bloodstream: the amount of carbon monoxide in a room and the length of exposure. Lower levels of CO inhalation can cause flu-like symptoms, including headaches, dizziness, disorientation, fatigue, and nausea. Other exposure effects can vary depending on the age and health of the individual.

Carbon monoxide in homes without gas appliances varies between 0.5 to 5 parts per million (ppm). CO levels in homes with properly maintained gas appliances will vary from 5 to 15 ppm. For those gas appliances that are not maintained properly, carbon monoxide levels may be 30 ppm or even higher.

The following authorities have set human exposure standards to carbon monoxide levels. The levels of CO that are measured in breathable air or air that is circulated in a home or business are:

- OSHA recommends a maximum of 50 ppm over 8 hours.
- EPA recommends a maximum level of 9 ppm over 24 hours.

Testing for Carbon Monoxide

Consumer products for detecting carbon monoxide in a home have been on the market for years. Carbon monoxide detectors should have an alarm that alerts consumers before they are exposed to hazardous levels of carbon monoxide. In order to prevent false alarms, CO detectors must be able to distinguish carbon monoxide gases from other types of gases, such as butane, heptane, alcohol, methane, and ethyl acetate. Two manufacturers that you can visit on the Internet to view the different types of carbon monoxide detectors available are *www.kidde.com* and *www.firstalert.com*.

Technicians who test for carbon monoxide use a special handheld meter to check the levels of carbon monoxide in a room or home. For a handheld carbon monoxide test meter, visit *www.fluke.com or www.fieldpiece.com*.

When to check for carbon monoxide in a home:

- When the consumer complains of headaches or nausea.
- Houseplants are dying.
- Unknown chronic odors from unknown sources.
- Condensation on cool surfaces that might lead to flue gas products in the home.

Figure 7-4 illustrates the locations in a home to test for carbon monoxide gas. These tests should be conducted near gas appliances, gas-heating systems, heating ducts, and the atmosphere in a room, approximately 6 ft above the floor.

When testing for CO in a gas appliance that has not been running for a while, you should follow these steps:

1. Test the air near the appliance and the surrounding air in the room before you turn on the appliance.
2. Test the air after you turn on the appliance.
3. Test the air near the appliance after the appliance has been running for 15 minutes.

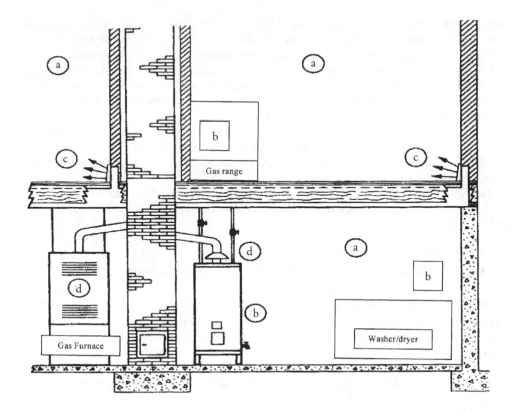

(a) Test in the atmosphere about 5 to 6 feet off the ground
(b) Test near gas appliances
(c) Test near heating ducts
(d) Test near appliance diverters and fire-doors on appliances in the basement or utility room

FIGURE 7-4   Typical test locations for carbon monoxide.

Flue Gases

Flue gases are the by-products of combustion that are exhausted through a chimney, or flue to the outside of the home. These gases consist of the following:

- *Nitrogen (colorless, odorless, and tasteless).* The air we breathe is made up of 79% nitrogen.

- *Carbon dioxide (colorless and odorless).* The human respiration process produces carbon dioxide.

- *Oxygen (colorless, odorless, and tasteless).* Oxygen is the main ingredient for combustion.

- *Carbon monoxide (colorless, odorless, and tasteless).*

- *Nitrogen oxide (colorless and odorless).* Nitrogen oxide forms in the combustion process of gas fuels.

- *Sulphur dioxide (colorless and smells like burnt matches).* Sulphur dioxide is irritating to the lungs.

- *Hydrocarbons (colorless, odorless, and tasteless).* Hydrocarbons are found in natural and liquefied petroleum gases.

- *Water vapor.* Approximately 10% will be vented through the flue or chimney.

- *Soot.* The remains of incomplete combustion.

The "Flame"

Prior to the invention of the Bunsen burner in 1842, gas burners would produce a yellow flame for light and heat. The process allowed the gas to enter into a tube, expel through a port, and, when lit with an ignition source, the gas would burn without pre-mixing the air before it left the burner (Figure 7-5). As the flame burned, the gas temperature began to rise

FIGURE 7-5
Gas lamp.

within the flame. Without the presence of pre-mixed air and gas, carbon particles began to pass through the flame, causing it to turn yellow. As more air was introduced into the combustion process, the flame would turn blue or blue with yellow tips.

Appliance manufacturers had the freedom to tailor the flame patterns from the Bunsen burner design to design gas appliances. The Bunsen burner worked on the principle of introducing air into the mixture of the gas before it left the burner port. An orifice was used to regulate the gas flow within the burner body (Figure 7-6). Attached to the burner body is an adjustable shutter to control the primary air mixture that enters into it. The flame that is produced by the Bunsen burner has multiple colors within it. Each color within the flame marks a stage of the burning process of gas (Figure 7-7). The stages are as follows:

- The inner cone is the first stage of the burning process. Within this cone, the gas is burned, which forms by-products, such as aldehydes, alcohols, carbon monoxide, and hydrogen.

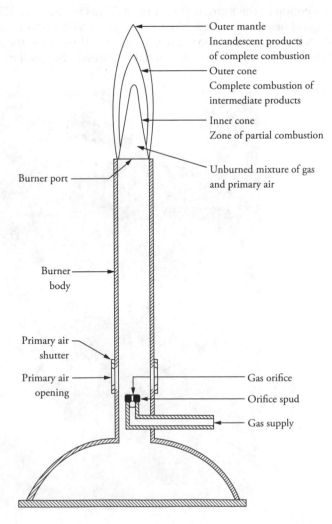

Figure 7-6
Bunsen burner.

Outer mantle
Incandescent products
of complete combustion

Outer cone
Complete combustion of
intermediate products

Inner cone
Zone of partial combustion

Unburned mixture of gas
and primary air

Burner port

Burner body

Primary air shutter

Primary air opening

Gas orifice

Orifice spud

Gas supply

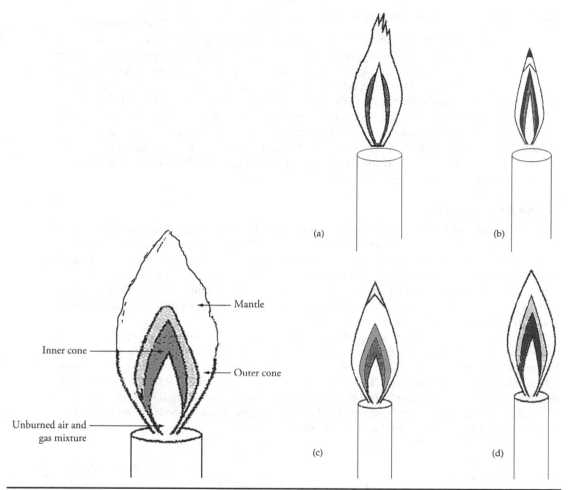

FIGURE 7-7 Different flame types of a Bunsen burner, depending on the amount of oxygen supplied.

- The outer cone surrounds the inner cone, and as air diffuses into the flame, it continues the burning process. If enough air is present during the burning process, the by-products from the inner cone will burn up in the outer cone. The by-products that are produced in the outer cone are carbon dioxide and water vapor.

- The outer mantle that is produced by the Bunsen burner is nearly invisible, with all the gas burned up. The reason it glows is due to the high temperature produced by the complete combustion of the gas products from the outer cone.

The temperature of a flame can vary, depending on what type of gas is used and how much air is pre-mixed with the gas. Temperatures within the flame will vary; the inner cone is cooler than the outer cone, and the hottest flame temperature is just above the outer cone.

Air Supply

When air is mixed with gas before it leaves the burner port, it is known as the primary air supply. Under ideal conditions, most burners only use 50% of the primary air supply to burn the gas, and the remaining 50% of the air is supplied by the secondary air supply from around the flames. In reality, additional excess air is needed to ensure that enough air is present to completely burn off the gas. The total air supply is accomplished by adding the sum of the primary, secondary, and excess air supplies. The total air supply needed for combustion is measured in percentages of air needed for complete combustion. Every manufacturer has its own specifications in the installation instructions or service manual for the total air percentages needed for the appliance model that you are servicing.

Flame Appearance

The appearance and stability of a flame is influenced by the amount of the primary air supply, the gas-burning speed, and the burner port. When the primary air supply is at 100%, under ideal conditions, the speed of burning is at its maximum. The speed of burning is also affected by the type of gas being burned. The flow velocity of the gas depends on the size of the orifice—the smaller the orifice size, the greater the flow velocity. Flow velocity from a burner port will also vary, depending on the size of the opening, with the greatest flow from the center of the port (Figure 7-8).

After the air-gas mixture leaves the port, the flow velocity slows down, and the flame begins to stabilize when the flow velocity equals the burning speed. At the same time, the flame cone begins to take shape when the flow velocity of the air-gas mixture levels off when it leaves the burner port. The burning speed increases and the flame temperature increases near the top of the flame cone. Therefore, the shape of the flame cone is rounded off at its tip. The inner cone is determined by the effects of the velocity away from the centerline of the port for each layer of the flame burning.

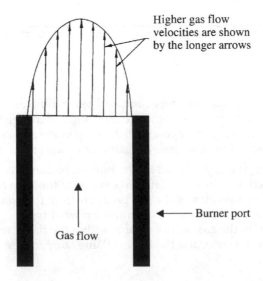

FIGURE 7-8
Flow velocities of gas.

Higher gas flow velocities are shown by the longer arrows

Burner port

Gas flow

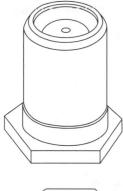

Figure 7-9 Gas
orifices are used to
limit gas flow to
burners.

Appliance manufacturers design gas appliances to have a stable burner flame by port loading. Port loading is expressed as BTU per hour per square inch of open burner port area. It is obtained by dividing the gas input rate by the total area of the port opening.

Gas Orifice

The gas orifice is not part of the burner, but it plays an important role in its operation (Figure 7-9). In the service field, the gas orifice goes by many names, such as orifice hood, spud, hood, or cap. The purpose of the gas orifice is to regulate or limit the flow of gas into the burner. The size of the hole in the orifice will depend on what type of gas is in use and what constant pressure is needed to light the burner evenly. Another reason for use of an orifice is to force primary air into the burner. Three types of orifices that are in use are:

1. The fixed orifice has a predetermined opening in the orifice hood to allow a certain rate of gas to flow (Figure 7-10).

2. The adjustable orifice is mainly used in ranges. This type of orifice will allow you to control the gas rate, from zero to its maximum rated flow.

3. The universal orifice is also mainly used on ranges. This type of orifice was designed to allow the range to be operated on natural or liquefied petroleum gas.

Most furnaces that are shipped from the factory are set up for natural gas operation. When the furnace needs to be converted to LP operation, the technician needs to purchase a conversion kit for the model being installed.

Figure 7-10
Different types of fixed orifices.

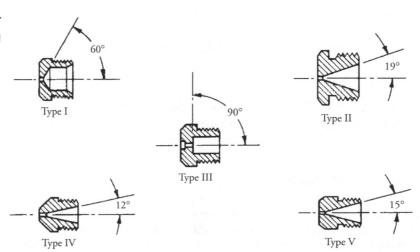

Ventilation

When you properly install a ventilation system, it will serve the following important purposes:

- The by-products of hot combustion gases will be transported outside the home.
- The home will be protected from fire hazards.
- A ventilation system will provide good air circulation and adequate oxygen supply for the gas appliance and the occupants of the residence.
- It will remove the water vapors produced when burning gas.

A wide variety of venting options are available from vent manufacturers, and every gas appliance manufacturer will provide venting instructions for properly installing their products. It is strongly recommended that you follow these instructions. You must use the proper venting materials as described in the installation instructions for the model being installed, and install the venting system according to your local building codes. For additional information on proper ventilation, visit *www.epa.gov*. Check out "Air for Combustion and Ventilation" sections of the National Fuel Gas Code, ANSI Z233.1 (latest edition) and check out NEPA31 and NEPA54 (National Fire Protection Association) for additional regulations. These codes are subject to change from time to time.

Troubleshooting Ventilation Problems

The most common problems with ventilation systems are:

- Improper maintenance of the ventilation system.
- Incorrect sizing and installation of the ventilation system.
- Inadequate air supply.

If you suspect a problem with the ventilation system, check the following:

- Make sure that you have the correct vent sizing.
- Check if there are too many elbows or if the length of the venting system exceeds the manufacturer's recommendations or local building codes.
- Inspect the entire ventilation system for faults, such as a disconnected or crushed pipe.
- Inspect for an obstruction, such as a clogged vent cap.
- Inspect all air openings for obstructions.

Measuring Gas Pressure

When installing or repairing gas appliances, it is often necessary to measure the pressure of the gas supply to the appliance. There also may be times when the technician will have to test the gas pressures at the manifold or orifices. The two types of test instruments used today to test the gas pressures are the manometer and the magnehelic gauge.

Figure 7-11
A manometer is used
to test gas pressure.

The manometer is a U-shaped tube equipped with a scale to measure gas pressure in water column inches (Figure 7-11). Before you check the gas pressure on an appliance, locate the model and serial number nameplate. This nameplate will have the gas rating stamped on it. Now turn off the gas supply to the appliance. Before you can begin to test the gas pressure, you must first set up the manometer by adding water to the U-shaped tube until both columns read zero on the scale (Figure 7-12a). If reading the scale is difficult, you can add a food coloring to the water to enhance the color so that you can read the scale better. When measuring gas pressure in an appliance, the tubing from the manometer is attached to the orifice, manifold, or gas supply line, and the other end of the tube on the manometer is left open to the room atmosphere. When the gas pressure is applied to the manometer, it pushes down on the water column and the water column rises on the other side of the manometer (Figure 7-12b). To read the manometer scale, observe where the water column stops. In Figure 7-12b, the incoming gas has pushed the water in the manometer down 2 in below zero, and the water column on the other side is pushed up 2 in above zero. The pressure reading is obtained by adding the two readings together. Figure 7-12b shows that the total change equals a 4-in water column. The reading should be within the rating on the model and serial nameplate.

The magnehelic gauge shown in Figure 7-13 also measures gas pressure. This gauge can measure gas pressure faster than the manometer can. To use this type of gauge, just follow the same procedures for the manometer. The only difference is that you read the gas pressure directly from the gauge scale. Some models have different scales on the gauge dial—just read the gauge that indicates water column pressure.

Many companies manufacture test instruments for checking gas pressures. Web sites to check out include:

- http://www.uniweld.com/catalog/gauges_thermometers/gas_pressure_test_kit.htm
- http://www.dwyer-inst.com
- http://www.marshbellofram.com

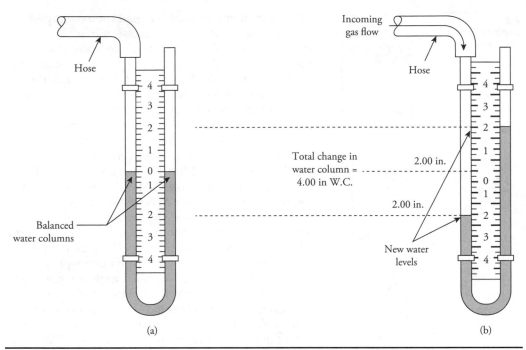

FIGURE 7-12 (*a*) Before using a manometer, check to see if the columns have equal amounts of water; and (*b*) when reading a manometer, you add the change in both columns; the sum equals the pressure of the gas.

FIGURE 7-13
A magnehelic gauge is used to test gas pressure.

Testing for Gas Leaks

When testing for a gas leak, do not use a lighter or a lit match. That is not the safe or correct way to test for gas leaks. All gas appliances, when they are installed or repaired, must be tested for leaks before placing the appliance in operation. Remember, LP or LPG is heavier than air and settles in low-lying areas. If you do not have a combustible gas detector (and I recommend that you purchase one), use your nose to smell for a gas leak all around the appliance at floor level. A safe and effective way to test for gas leaks is to use a chloride-free soap-and-water solution to check all connections and fittings. The solution can be sprayed, applied directly, or an applicator can be used to spread the solution. If a gas leak is detected by the soap solution it will begin to form bubbles.

Sometimes, air might be present in the gas lines from installation or repairs, and it could prevent the pilot or burner from lighting on initial startup. The gas lines should be purged of air in a well-vented area. Before proceeding with purging the gas lines, inspect the work area for any other sources of ignition, such as flames, burning candles, running electrical appliances, and the like. With a safe working environment, you can begin to purge the gas lines. On gas furnaces, turn off the gas supply valve as shown in Figure 7-14. Remove the gas line from the appliance, and slowly begin to turn on the gas supply valve. When you begin to smell gas, turn off the main supply valve. Next, reconnect the gas line to the furnace, open the gas supply valve, and test for gas leaks. If there are no gas leaks, run the furnace.

Use an approved pipe joint compound for connecting gas piping and fittings to the furnace. In addition, never use white Teflon tape on gas lines. White Teflon tape does not guarantee a leak-free connection. White Teflon tape is for water lines only. You can use a PTFE yellow Teflon tape on gas connections and fittings. This type of approved Teflon tape is a specially designed thread-sealing tape for use on gas appliances. To attach PTFE yellow Teflon tape to pipe thread, wrap the tape in the direction with the thread, as shown in Figure 7-15.

FIGURE 7-14
Main gas manual shut-off valve to the furnace. This shut-off valve must be located within six fcct of the furnaceThie supply line is a rigid pipe supply line. Pipe size of ¾" for natural gas or ½" for LP gas.

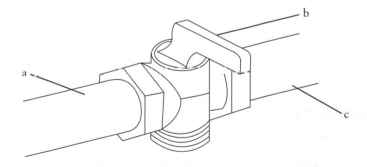

a. The gas supply line from the shut-off valve to the range
b. Gas supply-line shut-off valve
c. The main gas supply entering the home to the shut-off valve

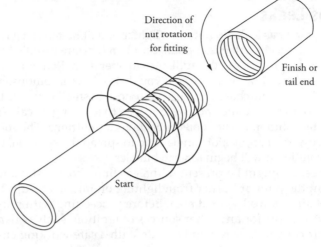

FIGURE 7-15
Wrap tape in the thread direction on a pipe fitting.

Direction of
nut rotation
for fitting

Finish or
tail end

Start

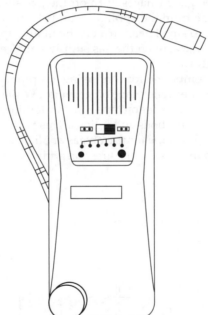

FIGURE 7-16
A combustible gas detector. This meter is used for detecting gas leaks.

Combustible Gas Detector

Combustible gas detectors are another general-purpose tool that the service technician could use to detect gas leaks (Figure 7-16). This instrument uses an audible alarm that gets louder when you approach a gas leak. This device has the capabilities of detecting LP (LPG) or natural gas and other hydrocarbons and halogenated hydrocarbons, depending on the manufacturer. The sensitivity of a combustible gas detector can be between 50 and 1,500 ppm of gas detection, with an instantaneous response.

One manufacturer of combustible gas detectors can be viewed on the Web at *www .aprobe.com*.

Oil

This chapter on oil is very important to the safety and technical knowledge that is needed to repair and maintain oil heating products. In addition, consumers should be knowledgeable of safety procedures and basic characteristics of fuel oil. Any person who cannot use basic tools or follow written instructions should not attempt to install, maintain, or repair oil-heating products.

If you do not fully understand the procedures in this chapter, or if you doubt your ability to complete the task on your oil product, please call your service company or service manager.

Crude oil is found underground in pools or pockets deep in the earth, and when the oil is extracted and refined it is then turned into "No. 2" fuel oil, which is used in oil burners for heating residential and light commercial properties. The designation of "No 2" fuel oil is used as a specification guide to define the physical characteristics of the oil, for example, flashpoint, viscosity, and the like. All fuel oils are not all alike, and its variations can have an impact on the burner operation.

Although this chapter cannot cover all there is to know about fuel oil, it will provide the basics.

Fuel Oil Safety

The following are a few safety tips to help you in handling oil heating appliances in your home or office:

- Always follow the manufacturer's installation instructions and the use and care manual for oil-heating appliances.

- Have a smoke detector and a carbon monoxide detector installed in the home or office, and check the batteries yearly.

- Teach your children not to play near or with your gas appliance.

- Check daily that the air-ventilation openings are clean and unobstructed. Also, check daily that nothing is blocking the burner inlet air openings.

- Check daily and make sure that combustible materials are not stored near the heating appliance.

- Check daily for signs of oil or water leaking around the burner or heating appliance. If leaking, contact a service company for repairs. Do not use the product until repairs are made.

- Check weekly the oil level in the storage tank, always keep it full. In the summer months, keep the storage tank full to prevent condensation or moisture from accumulating on the inside of the storage tank.

- Never attempt to burn any fuel not specified and approved by the manufacturer of the appliance.

- Do not operate the heating appliance beyond the specifications outlined in the manufacturer's use and care manual.

- To avoid electrical shock, disconnect the electrical supply before performing maintenance on the heating appliance.

- To avoid severe burns, allow the heating appliance to cool before performing maintenance.

- Do not operate any heating appliance if the heat exchanger is damaged, corroded, or pitted. Toxic flue products could enter the air stream and cause illness or death.

- Should overheating occur, do not turn off or disconnect the electrical supply to the heating appliance. Instead, shut off the oil supply at a location external to the appliance, if possible.

- Do not use a heating appliance if any part of it has been under water. Call a qualified service technician to inspect it.

- Do not operate a heating appliance if the appliance area will be exposed to air contaminants. See Tables 8-1 and 8-2.

Products to Avoid to Prevent the Potential of Severe Injury or Death When Operating Fuel Oil Appliances
Spray cans containing chloro/fluorocarbons
Permanent wave solutions
Chlorinated waxes/cleaners
Chlorine-based swimming-pool chemicals
Calcium chloride used for thawing
Sodium chloride used for water softening
Refrigerant leaks
Paint varnish removers
Hydrochloric acid/muriatic acid
Cements and glues
Antistatic fabric softeners used in clothes dryers
Chlorine-type bleaches, detergents, and cleaning solvents found in household laundry rooms
Adhesives used to fasten building products and other similar products

TABLE 8-1   Products That May Cause Contaminated Air for Oil-Heating Appliances

Areas Likely to Have Air Contaminants
Dry cleaning/laundry areas and establishments
Swimming pools
Metal-fabrication plants
Beauty shops
Refrigeration repair shops
Photo-processing plants
Auto body shops
Plastic manufacturing plants
Furniture refinishing areas and establishments
New building construction
Remodeling areas
Garages with workshops
Note: Refer to national, provincial or local codes for further information.

TABLE 8-2   For Safety, Remove the Contaminants Permanently or Isolate the Heating Appliance and Provide Outside Combustion Air.

Fuel Oil Ratings

When rating fuel oil there are standards and certain specifications that are set by the American Society for Testing and Materials (ASTM) Standards that must be considered:

- *Viscosity* is the oil's resistance to flow through an orifice within a specified amount of time. In other words, how well will the oil flow. Oil having a high viscosity will contribute to poor atomization of the oil, delayed flame ignition, noisy or pulsing flame, increased oil input, and the possibility of sooting, if the outside temperature is below 50°F. As the temperature drops, the viscosity increases.

- *Flash point* is the temperature at which the vapors escaping from the surface of the oil can be ignited and continues to burn. Depending which oil is used the flash point will be between 100 to 240°F (Table 8-1).

- *Pour point* is the temperature at which oil will barely flow. This is usually 5°F above the point when oil forms a solid mass. For No. 2 fuel oils, the ASTM D396 Standard for fuel oils lists the temperature at 20°F as the maximum pour point. To avoid cold climates, approximately 25% No. 1 oil (kerosene) is added to lower the pour point and cloud points.

- *Cloud point* is the temperature at which wax crystals begin to form. At around 10 to 20°F above the pour point, wax crystals form, which can clog filters and strainers, restricting the oil flow. By raising the oil temperature, the wax crystals will return to a solution.

- *Distillation temperature* of No. 2 oil is when the oil is vaporized and distilled (condenses) in a laboratory to determine the volatile components. At the moment when condensation begins it is called the initial boiling point (IBP).

- *Initial boiling point (IBP)*. The ignition source must provide enough heat energy to elevate the temperature of the atomized oil droplets to the initial boiling. If the initial boiling point is above 400°F, it could cause ignition problems.

The technician's first clue that the oil is not within ASTM specifications might be a sudden rash of malfunctions such as a delayed ignition of the oil, smoky fires, soot, noisy and/or dirty flames. Only a competent laboratory can determine if the oil is "out of spec." However, if the oil is within specifications, but is near the maximum level for viscosity, pour point, or has an IBP above 400°F, chemical additives and 25% kerosene might be considered to make the oil more compatible with colder temperatures and to improve ignition and combustion qualities.

Fuels and Their Flash Points

The flash point is an indication of how easy a chemical may burn. Materials with a higher flash point are less flammable or hazardous than chemicals with lower flash points. Table 8-3 illustrates the flash points of chemicals at atmospheric pressure.

Products of Refining Fuel Oil

One of the products extracted from crude oil is fuel oil. This process is done when crude oil is processed at the oil refinery. The products of refining are in Table 8-4.

Fuel Delivery, Storage, and Measurement

Fuel oil is delivered to the home or business by truck and pumped into storage tanks between 200 and 1,000 gal capacity. These storage tanks can be either stored above ground, below ground, or inside the building. Fuel oil is sold by the gallon and has a heating value of approximately 140,000 BTU/gal.

Advantages and Disadvantages of Underground Oil Tank Storage

The following are advantages and disadvantages of an oil storage tank underground installation (Figure 8-1).

Advantages:

- The storage tank does not take up any space in or out of the home.
- The storage tank is below the ground and it is out of sight.
- Large quantities of oil can be stored.
- Underground storage tanks are better insulated from the cold, compared with above ground storage tanks.

Disadvantages:

- The cost of installation, servicing, and inspection is too expensive.
- The storage tank is subject to cold and moisture intrusion.
- Today's storage tanks are well manufactured and they seldom leak. However, leak detection and clean-up are still important environmental concerns.

Fuel	Flash Point (°F)
Acetaldehyde	−36
Acetone	0
Benzene	12
Carbon disulfide	−22
Diesel fuel (1-D)	100
Diesel fuel (2-D0	125
Diesel fuel (4-D)	130
Ethyl alcohol	55
Fuel oil no. 1	100–162
Fuel oil no. 2	126–204
Fuel oil no. 4	142–240
Fuel oil no. 5	156–336
Fuel oil no. 6	150
Gasoline	−45
Gear oil	375–580
Iso-butane	−117
Kerosene	100–162
Methyl alcohol	52
n-Butane	−76
Propane	−156
Toluene	40
Xylene	63

To convert to Celsius use the following formula: C=5/9(°F)-32.

TABLE 8-3   The Flash Points of Different Chemicals

Advantages and Disadvantages of Above Ground Oil Storage Tanks

The following are advantages and disadvantages of an oil storage tank above ground installation (Figure 8-2).

Advantages:

- Less expensive to install and maintain.
- If leaks occur, they can be easily detected in time to avoid environmental problems.

Disadvantages:

- Storage tanks exposed to cold temperatures and moisture. The customer may have to provide a shelter for the tank.

Fuel Type	Used For
Gasoline	Used in combustion engines (automobiles and trucks)
Fuel oil no. 1	Kerosene, used for jet aircraft and as domestic fuel in small buildings
Fuel oil no. 2	Diesel oil, used for engines, identified by the addition of blue dye additive
Fuel oil no. 2	Oil used in heating systems, identified by the addition of red dye additive
Fuel oil no. 3	Oil used in heating systems requiring low viscosity fuel. ASTM merged this type oil fuel oil into fuel oil no. 2
Fuel oil no. 4	Oil used in small commercial and apartment buildings and is less expensive than fuel oil no. 2
Fuel oil no. 5	Oil used in industrial heating requiring it to be preheated first
Fuel oil no. 6	Oil used for fuel in large buildings and requires special equipment and preheating the oil before it can be used to burn properly

TABLE 8-4 Types of Fuel You Will Be Working With as a Service Technician Working on Oil Burners

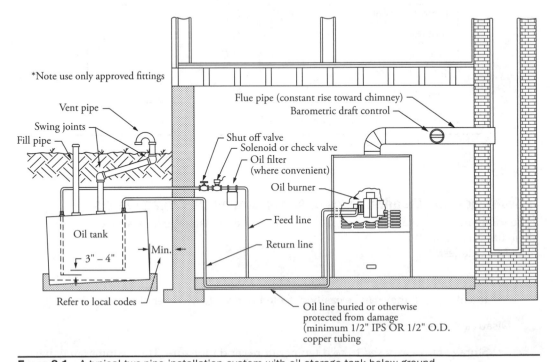

FIGURE 8-1 A typical two-pipe installation system with oil storage tank below ground.

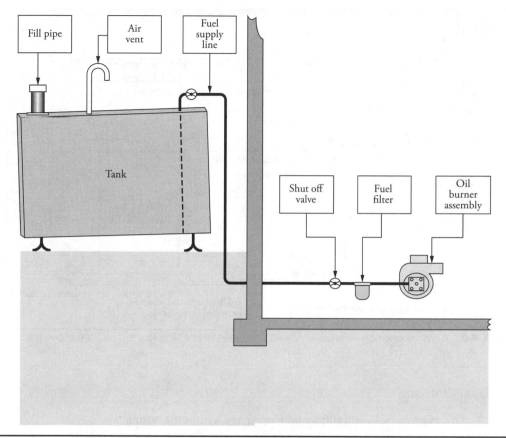

FIGURE 8-2 A typical one-pipe installation system with oil storage tank above ground.

- The storage tank takes up outdoor space.
- The storage tank may detract from the appearance of the home.

Advantages and Disadvantages of Indoor Oil Storage Tanks

The following are advantages and disadvantages of an oil storage tank installed indoors (Figure 8-3).

Advantages:

- The storage tank is not affected by outside cold temperatures and moisture.
- This type of tank installation is less expensive than underground storage tanks.
- Oil leaks are unlikely to occur. If they do, they are easily spotted and repaired.

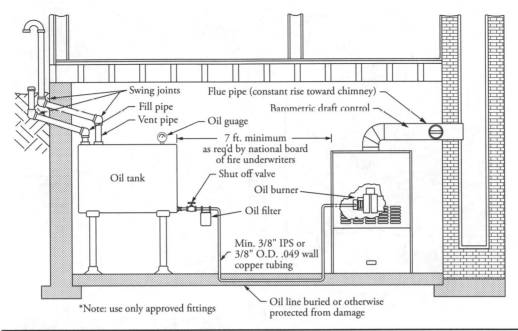

Swing joints

Flue pipe (constant rise toward chimney)

Fill pipe

Barometric draft control

Vent pipe — Oil guage

7 ft. minimum
as req'd by national board
of fire underwriters

Oil tank

Shut off valve

Oil burner

Oil filter

Min. 3/8" IPS or
3/8" O.D. .049 wall
copper tubing

*Note: use only approved fittings

Oil line buried or otherwise
protected from damage

FIGURE 8-3 A typical one-pipe installation system with oil storage tank located within the dwelling.

Disadvantages:

- This type of installation may take up space within the home.
- Some oil smell will be present within the home.

How the Fuel Oil Travels to the Furnace and Burner

The pumps, pipes, valves, filters, and controls are the components that transport the oil from the storage tank to the burner. The fuel-piping system is under negative pressure and relies on the fuel pump in the burner assembly to pull the oil from the storage tank to the burner. With the negative pressure in the oil fuel–piping system, any leaks will usually result in air being drawn into the fuel-supply system, potentially causing the system to become air-locked. For safety concerns, the components in the oil fuel supply must be free of leaks.

One-Pipe Fuel Oil Systems

When an oil storage tank is installed above ground, a single oil supply line will be installed from the oil storage tank to the burner assembly (Figures 8-2 and 8-3). This type of installation will ensure that the entire flow of fuel will be burned off in the burner. If there are any oil leaks in this type of system, the pump will lose its prime and become air-locked. When this happens, the pump must be bled and/or re-primed after the leaks have been repaired.

Two-Pipe Fuel Oil Systems

An oil storage tank that is installed in the ground will have one oil supply line and one return excess line installed leading back to the storage tank. With this type of system, the flow of fuel in the supply line is higher than what is used in the burner. The relief valve in the oil pump, which is set to regulate the burner's oil pressure, sends the excess oil back to the storage tank instead of internally relieving it to the outlet of the pump like a one pipe fuel oil system. This type of system is less susceptible to air-locking because of the higher volume of oil flow through the oil pump with a constant flow of fuel back to the storage tank.

Oil Filters

Oil filters (Figures 8-1 to 8-3) are used to filter the oil and are installed in the line between the storage tank and the pump. Furnace manufacturers recommend that the oil filter should have the capacity to trap a 40 to 50 µm particle. Also recommended is the use of a replaceable core, or spin-on type, supply line (suction) oil filter or dual filtration (primary and secondary) filters. As oil tanks age, sediment or sludge will accumulate. Without a filter, suspended particles of moisture and heavy oil that fall to the bottom of the storage tank can make their way into the fuel pump or nozzle. These oil filters should be replaced when the furnace has its routine maintenance performed. Typically there are only two to three different styles of oil filters. The technician should select the oil filter by gallon per hour (GPH) flow rate and pressure drop through the filter after determining the maximum flow rate, one-pipe or two-pipe system, requirements. The filter should be installed indoors and serviced/replaced annually. Oil filtration, single or dual, will protect the burner nozzle and prevent erratic or improper spray patterns that reduce efficiency, cause damage, higher maintenance costs, or the need for emergency service.

The Combustion Process

Fuel oil must be atomized into a fine mist and mixed with air before it can be ignited. When the fuel oil is burned, the chemical energy stored in the fuel oil is released into another form of energy called *heat*. To ignite the fuel oil droplets a higher heat source must be used. The electric spark delivered by the electrodes (Figure 8-4) of an oil burner provides the initial heat that causes oil droplets to become oil vapor and eventually burn evenly. This burning then heats the surrounding oil droplets causing them to burn also. This process continues until all of the oil droplets are vaporizing and burning. Under ideal conditions, all of the oil droplets will burn completely and cleanly within the combustion zone.

Fuel oil primarily consists of 85% carbon and 15% hydrogen. The combustion of fuel oil is the rapid combining of carbon, hydrogen, and oxygen, along with a source of ignition. The air we breathe consists of 79% nitrogen and 21% oxygen. The by-product of combustion besides the heat is carbon dioxide (CO_2), water vapor (H_2O), nitrogen oxides (NO_x), excess oxygen (O_2), sulfur oxides (SO_x), carbon (smoke), and carbon monoxide (CO). Figure 8-5 illustrates the combustion process under ideal conditions. Remember that air is primarily made up of nitrogen, which is not actually used in the combustion process but inherently present in the combustion air.

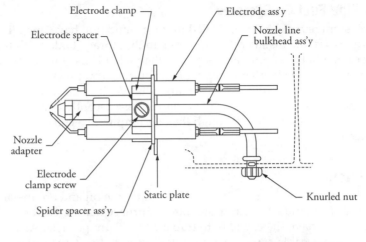

FIGURE 8-4
Typical gun assembly which houses the nozzle and the electrodes. The electrode will ignite the oil fuel and air mixture.

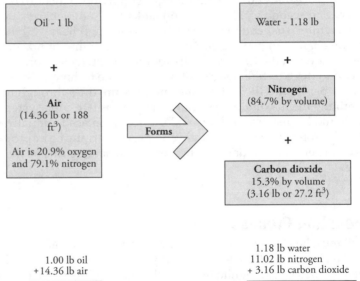

FIGURE 8-5
When 1 lb of fuel oil is burned (0% excess air) it will produce by weight and volume the following combustion products.

Excess Air

Excess air and combustion air are required to properly burn fuel oil. This excess air does not react during the combustion process but is a requirement to ensure the necessary mixing of fuel oil and oxygen. However, excess air is one of the major causes of lower furnace efficiencies. To understand how this can happen consider that excess air:

- Dilutes the combustion gases
- Absorbs heat
- Drops overall temperature of combustion gases

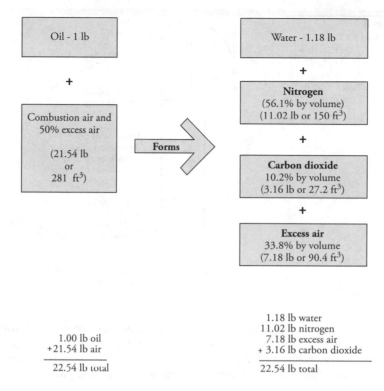

FIGURE 8-6
When 1 lb of fuel oil is burned (50% excess air) it will produce by weight and volume the following combustion products.

Oil - 1 lb

+

Combustion air and 50% excess air

(21.54 lb
or
281 ft³)

Forms →

Water - 1.18 lb

+

Nitrogen
(56.1% by volume)
(11.02 lb or 150 ft³)

+

Carbon dioxide
10.2% by volume
(3.16 lb or 27.2 ft³)

+

Excess air
33.8% by volume
(7.18 lb or 90.4 ft³)

1.00 lb oil
+21.54 lb air
———————
22.54 lb total

1.18 lb water
11.02 lb nitrogen
7.18 lb excess air
+ 3.16 lb carbon dioxide
————————————
22.54 lb total

- Lowers the flame temperature
- Causes a poorer heat exchange to the distribution medium

Excess air is the air that surrounds the flame. The additional oxygen that is provided by the excess air ensures that all of the carbon and hydrogen in the fuel oil comes in contact with ample oxygen to properly burn. Figure 8-6, illustrates the combustion process with the additional 50% excess air.

The Relationship of Excess Air, Smoke, and Efficiency

During the combustion of fuel oil, some smoke is generated because some of the oil droplets do not contact enough oxygen to complete the reaction that forms carbon dioxide. This smoke consists mainly of small particles of unburned carbon. As the manufacturers have learned, as the amount of excess air is increased, the efficiency of transferring heat to the heat exchange medium (hot water, warm air, or steam) is reduced. This increase of excess air also reduces the amount of smoke. However, as the amount of excess air is reduced, both the amount of smoke and the efficiency of the system will increase. While the smoke is emitted through the stack, some of the smoke particles will stick to the inside of the heat exchanger surface and acts as insulation and can eventually clog up the flue passages. Over a period of time this build up, called "sooting" leads to reduced efficiency and more maintenance and service calls.

Figure 8-7
A typical smoke and efficiency chart. This chart will vary from manufacturer to manufacturer and will vary with different burners.

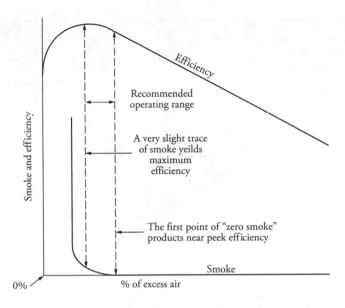

There is a delicate balance between the combustion process and the smoke generation caused by insufficient excess air and the reduced heat-transfer efficiency and an increased volume of combustion products caused by too much excess air. Figure 8-7 illustrates a typical chart that compares smoke and efficiency and will vary from manufacturer to manufacturer.

BTU Value of Fuel Oil versus Gas Fuel

Listed in Table 8-5 is the BTU (British Thermal Unit—common measurement used to define the quantity of heat being moved) rating for fuel oils and gases.

Type	Unit	BTU
Fuel oil no. 1	Gallon	137,400
Fuel oil no. 2	Gallon	139,600
Fuel oil no. 3	Gallon	141,800
Fuel oil no. 4	Gallon	145,100
Fuel oil no. 5	Gallon	148,800
Fuel oil no. 6	Gallon	152,400
Natural gas	Cubic Foot	950–1,050
Propane	Cubic Foot	2,550
Butane	Cubic Foot	3,200

Table 8-5 BTU Value of Fuel Oil versus Gas Fuel (1 ft^3 = 7.480519 gal)

General Venting Requirements

Failure to follow the manufacturer's recommendation on proper venting could result in flue gas spillage and carbon monoxide emissions, causing severe personal injury or death. The chimney must be inspected and cleaned before any new installation. Replacement or repairs must be completed on the existing chimney if a visual inspection indicates the chimney may be unusable for use. Insufficient draft can cause flue gas leakage and carbon monoxide emissions into the residence or work area.

Venting of heating appliances should be limited to the outside and in accordance with local codes or requirements of local utility. For additional information, refer to ANSI/NEPA 211 Chimney, Fireplaces, Vents, and Solid Fuel Burning Appliances (*www.nfpa.org*) and/or CSA B139 Installation Code (*www.mha-net.org*).

General Venting Requirements

Room Air Conditioners

Many excellent books have been written on the subject of air conditioning and refrigeration, and this chapter will only cover the basics needed to perform preventive maintenance on room air conditioners. Each preventive maintenance procedure has been rated by a degree of complexity or simplicity. For the owner of the product, you will have to decide whether you can perform the preventive maintenance procedure or if you must call a professional to complete the preventive maintenance procedures on the product. The chapter will not cover diagnostics, any electrical repairs, or replacement procedures of any sealed system components. This is a specialized area, and only an EPA-certified technician can repair the sealed system in room air conditioners. The certified technician has the proper training, tools, and equipment required to make the necessary repairs to the sealed system. For the new HVAC student, the information in this chapter will be of great importance in your studies to become a certified air conditioning technician. An individual who is not certified to service the sealed system could raise the risk of personal injury, as well as property damage. The uncertified individual could void the manufacturer's warranty on the product if he or she attempts entry into the sealed system. Following are the symbols that are used in this chapter to define who should perform the preventive maintenance on the room air conditioner.

Expert		Call a certified HVAC service Company
Hard		Service technician
Moderate		Business Owner
Easy		Home Owner

Principles of Operation

The room air conditioner, when installed and running properly, will circulate the air in a room or area, removing the heat and humidity; some models will heat the air in the winter months. At the same time, the air filter, located behind the front grille, will filter out dust particles. Most models have a fresh air intake feature, which allows fresh outside air to enter the room when the unit is running. The thermostat will control the comfort level in the room or area, and cycle the air conditioner on and off according to the temperature setting.

Safety First

Any person who cannot use basic tools or follow written instructions should *not* attempt to install, maintain, or repair any room air conditioners. Any improper installation, preventive maintenance, or repairs could create a risk of personal injury or property damage. If you do not fully understand the preventive maintenance procedures in this chapter, or if you doubt your ability to complete the task on your room air conditioner, please call your HVAC service professional.

This chapter covers the preventive maintenance procedures and does not cover how to diagnose the sealed system. The actual repair or replacement of any sealed-system component is not included in this chapter. It is recommended that you acquire refrigerant certification (or call an authorized service company) to repair or replace any sealed-system component, as the refrigerant in the sealed system must be recovered properly.

Before continuing, take a moment to refresh your memory on the safety procedures in Chapter 1.

Room Air Conditioners in General

Much of the preventive maintenance information in this chapter covers room air conditioners in general, rather than specific models, in order to present a broad overview of preventive maintenance techniques. The illustrations that are used in this chapter are for demonstration purposes only, to clarify the description of how to perform preventive maintenance on these appliances. They in no way reflect on a particular brand's reliability.

Room Air Conditioner Preventive Maintenance

Room air conditioners (including portable models) have air filters (Figure 9-1) that need cleaning every 225 to 360 fan hours of operation. The discharge grille area also needs vacuuming to remove the dust buildup. Twice a year, the following areas need to be inspected and/or cleaned:

- The evaporator coil
- The condenser coil
- The evaporator pan and base pan
- The indoor blower housing and blower wheel
- All the wiring connections and wiring
- The electrical and mechanical controls
- The voltage at the receptacle
- The inside and outside of the air conditioner
- All gaskets
- The drain system (clean it, too)
- The cabinet seal (clean the outer cabinet)
- The copper tubing

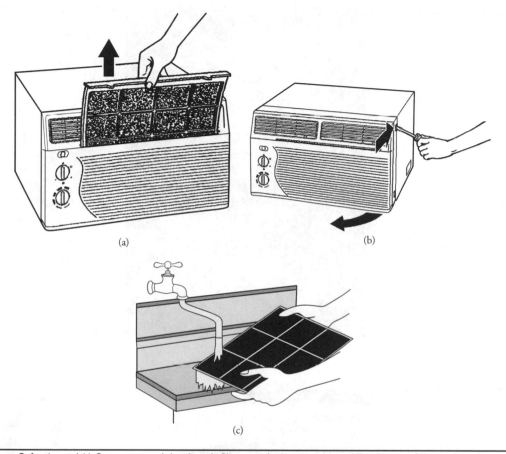

FIGURE 9-1 (*a* and *b*) On some models, the air filter can be removed by sliding it out of the unit or by removing the front grille. Be careful not to damage the plastic tabs on the grille; (*c*) wash filter with warm water and shake excess water out of the filter.

Always check the manufacturer's literature that comes with the room air conditioner if there are additional preventive maintenance checks that are not listed on this checklist.

Inspecting Room Air Conditioners

Twice a year you must inspect all control components: electrical and mechanical, electronic, as well as the power supply and the remote controller (if a controller comes with the model you are servicing) (Figure 9-2*a* and *b*). Also, inspect the air conditioner plug and receptacle for signs of damage or burning. The technician must use the proper testing instruments (voltmeter, ohmmeter, ammeter, and the like) to perform electrical tests (Figure 9-3*a* and *b*). The technician should also use a thermometer (Figure 9-3*c* and *d*) to test the indoor and the outdoor coil operating temperatures. Use a psychrometer (Figure 9-4) to measure the wet bulb and dry bulb temperatures indoors and outdoors.

FIGURE 9-2
(a) Inspect all electrical components in the control panel; (b) remove the cover, inspect the batteries, and re-attach the cover. Replace the batteries if the controller does not respond to commands;

(a)

(b)

FIGURE 9-3
(a) Multimeter, (b) clamp-on ammeter, (c) IR thermometer, and (d) analog thermometer. These test instruments are used for testing room air conditioners. (*Continued*)

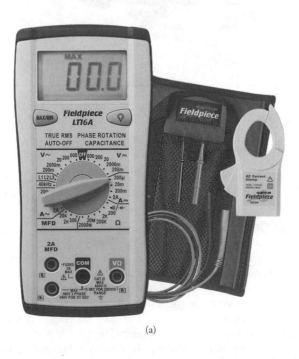

(a)

(b)

FIGURE 9-3
(*a*) Multimeter,
(*b*) clamp-on ammeter,
(*c*) IR thermometer,
and (*d*) analog
thermometer. These
test instruments are
used for testing room
air conditioners.

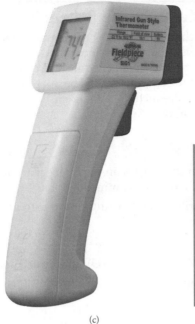

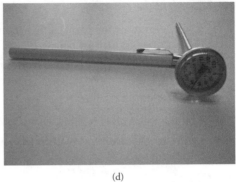

(c)

(d)

FIGURE 9-4
A handheld
psychrometer.

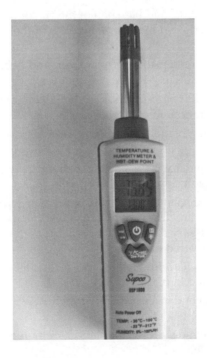

Room Air Conditioner Preventive Maintenance Schedule

The following preventive maintenance schedule will assist the homeowners, business owners, and service technicians set a preventive maintenance schedule for room air conditioners. By sticking to a preventive maintenance schedule and properly maintaining the room air conditioner, the efficiency of the equipment will be operating at or above the 90% range of the rated BTU capacity with a reduction in energy use. Following activities are suggested to maintain the air conditioner at optimal level:

 Monthly

- Check filters for particulate accumulation. Clean or replace if accumulation results in a drop in airflow below manufacturers recommendations.
- Check the integrity of the front grille, filter housing and filter, correct as needed.

 Quarterly

- If the room air conditioner has a UV lamp, check the UV lamp, clean or replace as needed.
- Check the drain pan, drain line, and base pan of the air conditioner for proper water drainage.

 Semi-annually

- Check air conditioner for proper operation. Repair, adjust, or replace defective components to ensure proper operation.
- Check the drain pan, drain line, and base pan of the air conditioner for proper water drainage. Remove any debris that may block the flow of the condensate water. Clean as needed.
- Check the evaporator and condenser coil for any dirt and debris. Clean, restore, or replace as needed.
- Check the control box and its components including the wiring and wiring connectors for proper operation. Repair or replace as needed.
- Check blower wheel and fan blade. Clean, repair, or replace as needed.
- The temperature differential of the intake and discharge air through the evaporator coil. If the temperature difference is outside the manufacturer's recommendations, find the cause, repair and adjust the refrigerant levels.

  Note: An EPA-certified technician only can repair the sealed system in room air conditioners. Homeowners and business owners must call a certified service company to make the necessary sealed system repairs, as the refrigerant in the sealed system must be recovered properly.

- Check the integrity of the air conditioner. Make sure that all of the screws, fasteners, and brackets are secured and in place.
- Check the evaporator and condenser coil fins, straighten as needed. This will help with increased air flow.
- Check areas that may have moisture accumulation for biological growth. If present, clean and disinfect as needed.

Cleaning a Room Air Conditioner

When cleaning the room air conditioner, use an approved cleaner to wash the unit. Remember to protect the electrical components and fan motor with plastic to prevent the water from damaging the components. Refer to the use and care manual that comes with every air conditioner for further maintenance instructions on the model you are servicing. Do not plug in or run the air conditioner after using water to clean the unit. Wait a few hours, allowing the air conditioner to completely dry out. To prevent electrical mishaps, the air conditioner must be totally dry before you can plug it in.

Prepare and Clean the Room Air Conditioner

Before you begin to wash the air conditioner, you must disassemble the unit to gain excess to the coils and the base pan. (See Figures 9-5.)

After disassembly (Figure 9-6), take a plastic bag and cover the fan motor and control box (Figure 9-7a and b). Place the unit in an area to be cleaned. Spray a non-corrosive

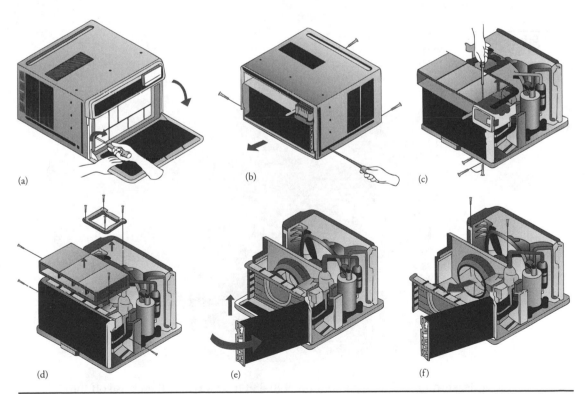

FIGURE 9-5   (a) Remove the front grille; (b) remove screws that secure the air conditioner chassis to the outer case; (c) remove the screws that secure the upper evaporator fan shroud and control box; (d) remove the bracket that secures the fan shroud and separate the fan shroud from the air conditioner; (e) remove screws that secure evaporator coil and swing the evaporator coil away from the base slightly; (f) remove the evaporator blower wheel; (g) remove the screws that secure the condenser coil to the air conditioner chassis. Then swing the condenser coil away from the unit slightly; and (h) remove condenser fan blade and shroud. (*continued*)

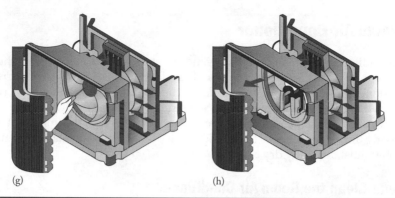

(g)   (h)

FIGURE 9-5 (*a*) Remove the front grille; (*b*) remove screws that secure the air conditioner chassis to the outer case; (*c*) remove the screws that secure the upper evaporator fan shroud and control box; (*d*) remove the bracket that secures the fan shroud and separate the fan shroud from the air conditioner; (*e*) remove screws that secure evaporator coil and swing the evaporator coil away from the base slightly; (*f*) remove the evaporator blower wheel; (*g*) remove the screws that secure the condenser coil to the air conditioner chassis. Then swing the condenser coil away from the unit slightly; and (*h*) remove condenser fan blade and shroud.

FIGURE 9-6
Room air conditioner disassembled and ready for cleaning.

cleaner on the coils, shroud, and base pan (Figure 9-8). Let it set a while, then rinse off the coils, shroud, blower fan, and base pan (Figure 9-9). Also, clean the front grille and cabinet parts. Let the water drain from the base pan. Inspect the compressor for any rust or corrosion (Figure 9-10). Remove the plastic from the control box and from the fan motor. Next, start reassembling the air conditioner in the reverse order of disassembly. Let the air conditioner dry for a few hours before turning the unit on for testing. Finally, reinstall the

(a) (b)

FIGURE 9-7 (a) Before washing air conditioner, wrap fan motor in plastic to prevent water from entering the motor; (b) wrap the control box in plastic to protect the controls from water.

FIGURE 9-8
Spray non-corrosive cleaning solution on shroud, coils, and base pan.

air conditioner back in its place. Follow the installation instructions for the proper way to re-install the air conditioner.

Performance Data

After you have completed the maintenance on the air conditioner, perform an electrical test by checking the amperage or wattage on the unit, and compare the readings with the information on the model number data tag. At the same time, perform readings on the following:

- The room temperature and outside temperature.

- The temperature differential of the intake and discharge air through the evaporator coil. Take a reliable thermometer, place it in front of the air intake (where the air filter

(a)   (b)

FIGURE 9-9   Rinse of chemicals from coils, blower fan, and base pan. Make sure not to get water on the control box and fan motor. Let air conditioner dry before re-assembly. Do not plug in unit; wait a few hours for air conditioner to dry completely.

FIGURE 9-10
If compressor is rusted and leaking oil, replace room air conditioner if it is out of warranty.

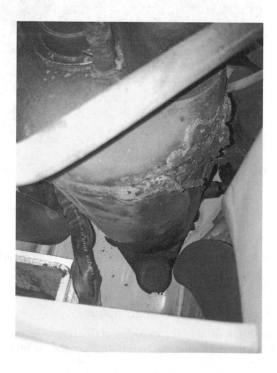

is located), and take a reading. Then place the thermometer in the discharge grille, and take a reading of the air blowing into the room. The difference between the two readings will be the temperature drop. This reading will vary among manufacturers and models. The temperature drop should be between 18° and 31°F.

- The temperature differential of the intake and discharge air through the condenser coil. Use the same reliable thermometer to take the readings.

- Use a psychrometer to measure the indoor and the outdoor wet bulb temperatures. The sling psychrometer will measure the relative humidity in the room and outside (Table 9-1).

Temperature of the Room or Outside (Dry Bulb) in Degrees Fahrenheit	The Temperature Difference between the Dry Bulb and the Wet Bulb Temperatures in Degrees Fahrenheit								
	4	5	6	7	8	9	10	11	12
	Relative Humidity % at Pressure = 30.00 in								
40	68	60	52	45	37	29	22	15	7
50	74	67	61	55	49	43	38	32	27
52	75	69	63	57	51	46	40	35	29
54	76	70	64	59	53	48	42	37	32
56	76	71	65	60	55	50	44	39	34
58	77	72	66	61	56	51	46	41	37
60	78	73	68	63	58	53	48	43	39
62	79	74	69	64	59	54	50	45	41
64	79	74	70	65	60	56	51	47	43
66	80	75	71	66	61	57	53	48	44
68	80	76	71	67	62	58	54	50	46
70	81	77	72	68	64	59	55	51	48
72	82	77	73	69	65	61	57	53	49
74	82	78	74	69	65	61	58	54	50
76	82	78	74	70	66	62	59	55	51
78	83	79	75	71	67	63	60	56	53
80	83	79	75	72	68	64	61	57	54
82	84	80	76	72	69	65	61	58	55
84	84	80	76	73	69	66	62	59	56
86	84	81	77	73	70	66	63	60	57
88	85	81	77	74	70	67	64	61	57
90	85	81	78	74	71	68	65	61	58
100	86	83	80	77	73	70	68	65	62

A wet bulb depression chart. Formula: DB −WB = WBD.

TABLE 9-1   Psychrometric Table

- Measure the operating voltage.
- Measure the startup and cycling amperage or wattage of the unit.

Take the readings and match them against the manufacturer's performance data. You can locate the air conditioner performance data on the manufacturer's Web site or in the manufacturer's service manual. The data that you accumulated should match—if it doesn't, adjustments will have to be made to bring the air conditioner up to manufacturer's standards. You might have to replace a component, clean the unit, or correct the installation.

10 CHAPTER

Residential Air Conditioning and Heating

This chapter will only cover the basics needed to perform the preventive maintenance on residential air conditioning and heating systems. Each preventive maintenance procedure has been rated by a degree of complexity or simplicity. For the owner of the product, you will have to decide whether you can perform the preventive maintenance procedure or if you must call a professional to complete the preventive maintenance procedures on the product. Following are the symbols that are used in this chapter to define who should perform the preventive maintenance on the air conditioning and heating equipment.

The home owner and/or business owner will need hand tools and a flashlight or drop light to gain access and to see into the HVAC equipment.

The chapter will not cover diagnostics, any electrical repairs, or replacement procedures of any sealed system components. This is a specialized area, and only an EPA-certified technician can repair the sealed system on air conditioning equipment. The certified technician has the proper training, tools, and equipment required to make the necessary repairs to the sealed system. For the new HVAC student, the information in this chapter will be of great importance in your studies to become a certified air conditioning technician. An individual who is not certified to service the sealed system could raise the risk of personal injury, as well as property damage. The uncertified individual could void the manufacturer's warranty on the product if he or she attempts entry into the sealed system.

Air Conditioning and Heating Equipment

The air conditioning equipment involves many aspects of conditioning the air in a residence and/or including introducing fresh air in whatever way necessary to make the living environment for the occupants comfortable. This process may include cooling the air, warming the air, filtering the air, adding moisture to the air, removing moisture from the air, and maintaining a well-balanced air flow and duct distribution system. Air conditioning includes:

- Cooling the occupied space
- Heating the occupied space
- Humidification (moisture in the air)
- Dehumidification (removing moisture from the air)
- Preventive maintenance on the equipment
- Cleaning the equipment
- Filtration of the recirculating air within the conditioned space
- Humidity control within the occupied space

Principles of Operation

The residential air conditioning unit, when installed and running properly, will circulate the air through ducts in the room(s) or area, removing the heat and humidity; some models will heat the air by electric, gas, oil, or water in the winter months. At the same time, the air filter, located in the return air grille, will filter out dust particles. Some models have a fresh air intake feature, which allows fresh outside air to enter the return duct when the unit is running. The thermostat will control the comfort level in the room(s) or area, and cycle the air conditioner on and off according to the temperature setting.

Safety First

Any person who cannot use basic tools or follow written instructions should *not* attempt to install, maintain, or repair any residential air conditioning unit. Any improper installation, preventive maintenance, or repairs could create a risk of personal injury or property damage. If you do not fully understand the preventive maintenance procedures in this chapter, or if you doubt your ability to complete the task on your residential air conditioner, please call your HVAC service professional.

This chapter covers the preventive maintenance procedures and does not cover how to diagnose the sealed system. The actual repair or replacement of any sealed-system component is not included in this chapter. It is recommended that you acquire refrigerant certification (or call an authorized service company) to repair or replace any sealed-system component, as the refrigerant in the sealed system must be recovered properly.

Before continuing, take a moment to refresh your memory on the safety procedures in Chapter 1.

Residential Air Conditioning and Heating Equipment in General

Much of the preventive maintenance information in this chapter covers residential air conditioners in general, rather than specific models, in order to present a broad overview of preventive maintenance techniques. The illustrations that are used in this chapter are for demonstration purposes only, to clarify the description of how to perform preventive maintenance on these appliances. They in no way reflect on a particular brand's reliability.

Preventive Maintenance Checklist

The following checklist will assist the certified HVAC service technician in performing a proper inspection on residential air conditioning and heating equipment. By properly maintaining the equipment, the efficiency of the equipment will be operating at or above the 90% range of the rated BTU capacity with a reduction in energy use. Always check the manufacturer's literature that comes with the air conditioning and heating equipment if there are additional preventive maintenance checks that are not listed on this checklist.

Central Air Conditioning Unit

- Check and adjust the thermostat.
- Check to make sure indoor and outdoor unit turn on.
- Check sequence of operation.
- Replace air filter or clean reusable type filter.
- Check and inspect the ducts to make sure that there are no air leaks or discrepancies.
- Check the duct grilles and registers for proper air flow.
- Check the evaporator coil and advise if dirty or if it needs cleaning.
- Check expansion valve and coil temperatures.
- If the unit you are inspecting has a belt-driven blower motor, check the blower belt wear, tension, and adjust as needed.
- Check bearings and lubricate blower motor if needed.
- Check the condensate drain system and secondary drain pan then advise of any discrepancies.
- Check condensate drain pan float switches for proper operation and installation.
- Check the condenser coil to determine if it needs cleaning.
- Inspect all wiring and connections to controls and electrical connections.
- Check voltage and amperage draw on all motors with a clamp-on multimeter.
- Check compressor contactor, capacitor, and any other controls in the condenser unit.
- Check the compressor and the amperage draw and compare the reading against the data information plate that is attached to the unit.
- If the unit you are inspecting has a start capacitor and potential relay, check it.
- If the unit you are inspecting has a pressure switch, check the cut-out settings.
- Install refrigerant gauges and check operating pressures, superheat, and subcooling.
- Check refrigerant levels and advise the owner if adjustment or repairs are necessary.
- Check the conditions of the entire air conditioning system and advise the owner of any discrepancies.

Electric Heat

- Check and adjust the thermostat.
- Check and see if the indoor unit turns on.
- Check sequence of operation.
- Replace air filter or clean reusable type filter.
- Check and inspect the ducts to make sure that there are no air leaks or discrepancies.
- Check the duct grilles and registers for proper air flow.
- Check the evaporator coil and advise if dirty or if it needs cleaning.
- If the unit you are inspecting has a belt driven motor, check the blower belt wear, tension, and adjust as needed.
- Check bearings and lubricate blower motor if needed.
- Check voltages to the unit and check the voltage in the control box.
- Check the amperage of the blower motor and compare reading against the data information plate that is attached to the unit.
- Check the amperage of each electric heating element and compare against the data information plate that is attached to the unit.
- Check the total amperage draw for all of the electric heating elements and compare the reading against the data information plate that is attached to the unit.
- Check all of the heat sequencers for proper operation.
- Check electrical wiring and connections.
- Check the temperature rise.
- Check the supply temperature.
- Check the heat anticipator on the thermostat for the proper setting.
- Check the conditions of the entire air conditioning system and advise the owner of any discrepancies.

Heat Pump

- Check and adjust thermostat.
- Check and make sure indoor and outdoor units turn on.
- Check sequence of operation.
- Replace air filter or clean reusable type filter.
- Check and inspect the ducts to make sure that there are no air leaks or discrepancies.
- Check the duct grilles and registers for proper air flow.
- Check the evaporator coil and condenser coil and advise if dirty or if it needs cleaning.
- Check the condensate drain system and secondary drain pan then advise of any discrepancies.
- Check condensate drain pan float switches for proper operation and installation.

- Check expansion valve and coil temperatures.
- If the unit you are inspecting has a belt-driven motor, check the blower belt wear, tension, and adjust as needed.
- Check bearings and lubricate blower motor if needed.
- Check electrical wiring and connections.
- Check voltages to the unit and check the voltage in the control box.
- Check the blower motor amperage draw and compare the reading against the data information plate that is attached to the unit.
- Check the amperage of each electric heating element and compare against the data information plate that is attached to the unit.
- Check the total amperage draw for all of the electric heating elements and compare the reading against the data information plate that is attached to the unit.
- Check all of the heat sequencers for proper operation.
- Check the condenser fan motor bearings and lubricate if needed.
- Check the condenser motor amperage draw and compare the reading against the data information plate that is attached to the unit.
- Check compressor contactor, capacitor, and any other controls in the condenser unit.
- Check voltage and amperage draw on the compressor with a clamp-on multimeter draw and compare the reading against the data information plate that is attached to the unit.
- Check the crankcase heater on the compressor if one is installed.
- Check the defrost controls for proper operation.
- Check reversing valve for proper operation.
- If the unit you are inspecting has a start capacitor and potential relay, check it.
- If the unit you are inspecting has pressure switches, check for proper operation.
- Install refrigerant gauges and check operating pressures, superheat, and subcooling.
- Check refrigerant levels and advise the owner if adjustment or repairs are necessary.
- Check the conditions of the entire heat-pump system and advise the owner of any discrepancies.

Gas Furnace

- Check and adjust the thermostat.
- Check the heat anticipator.
- Check and see if the gas furnace comes on.
- Check sequence of operation.
- Replace air filter or clean reusable type filter.
- Check condensate drainage system and secondary drain pan and advise of any discrepancies.
- For gas furnaces with cooling coil, check condensate drain pan(s) and float switches for proper operation and installation.

- Check and inspect the air ducts and to make sure that there is no air leaks or discrepancies.
- Check the duct grilles and registers for proper air flow.
- If the gas furnace you are inspecting has a belt-driven blower motor, check the blower belt wear, tension, and adjust as needed.
- Check bearings and lubricate blower motor if needed.
- Check voltages to the unit and check the voltage in the control box.
- Check the amperage of the blower motor and compare reading against the data information plate that is attached to the unit.
- Check the flue for rust and corrosion and advise of discrepancies.
- Check, clean, and adjust pilot if needed.
- Check electronic spark ignition control for proper operation.
- Check all wiring and connections to controls and electrical connections.
- Check incoming gas pressures and leaving gas pressures from gas valve manifold.
- Check burners to see if they need cleaning and advise.
- Check and adjust burners for fuel efficiency.
- Check heat exchanger for cracks, soot, and rust.
- Check the heat exchanger for cracks when the furnace is hot.
- Check blower motor and induce draft motor amperage and compare reading against the data information plate that is attached to the unit.
- Check fan controls for proper operation.
- Test safety shutoff response.
- Check the conditions of the entire gas furnace system and advise the owner of any discrepancies.
- Check the CO_2 monitor and the smoke alarm for proper operation; replace batteries yearly.

Dual Fuel or Heat Pump Furnace

- Check and adjust the thermostat.
- Check sequence of operation.
- Check to make sure indoor and outdoor unit turn on, including the furnace.
- Replace air filter or clean reusable type filter.
- Check and inspect the ducts to make sure that there are no air leaks or discrepancies.
- Check the duct grilles and registers for proper air flow.
- Check the evaporator coil and advise if dirty or if it needs cleaning.
- Check the condensate drain system and secondary drain pan then advise of any discrepancies.

- Check condensate drain pan float switches for proper operation and installation.
- Check expansion valve and coil temperatures.
- If the unit you are inspecting has a belt-driven motor, check the blower belt wear, tension, and adjust as needed.
- Check bearings and lubricate blower motor if needed.
- Check electrical wiring and connections.
- Check voltages to the unit and check the voltage in the control box.
- Check the blower motor amperage draw and compare the reading against the data information plate that is attached to the unit.
- Check the amperage of each electric heating element and compare against the data information plate that is attached to the unit.
- Check all of the heat sequencers for proper operation.
- Check the condenser fan motor bearings and lubricate if needed.
- Check the condenser motor amperage draw and compare the reading against the data information plate that is attached to the unit.
- Check compressor contactor, capacitor, and any other controls in the condenser unit.
- Check voltage and amperage draw on the compressor with a clamp-on multimeter; draw and compare the reading against the data information plate that is attached to the unit.
- Check the crankcase heater on the compressor if one is installed.
- Check the defrost controls for proper operation.
- Check reversing valve for proper operation.
- If the unit you are inspecting has a start capacitor and potential relay, check it.
- If the unit you are inspecting has pressure switches, check for proper operation.
- Check the flue for rust and corrosion and advise of discrepancies.
- Check the flue for proper operation.
- Check, clean, and adjust pilot if needed.
- Check electronic spark ignition control for proper operation.
- Check and adjust burners for fuel efficiency.
- Check heat exchanger for cracks, soot, and rust.
- Check the heat exchanger for cracks when the furnace is hot.
- Check blower motor and induce draft motor amperage and compare reading against the data information plate that is attached to the unit.
- Check incoming gas pressures and leaving gas pressures from gas valve manifold.
- Check fan controls for proper operation.
- Test safety shutoff response.
- If the unit you are inspecting has a pressure switch, check the cut-out settings.

- Install refrigerant gauges and check operating pressures, superheat, and subcooling.
- Check refrigerant levels and advise the owner if adjustment or repairs are necessary.
- Check the conditions of the entire system and advise the owner of any discrepancies.
- Check the CO_2 monitor and the smoke alarm for proper operation, replace batteries yearly.

Oil Furnace

- Check and adjust the thermostat.
- Check the heat anticipator.
- Check sequence of operation.
- Check and see if the oil furnace comes on.
- Check flame characteristics.
- Check temperature rise.
- Replace air filter or clean reusable type filter.
- Check and inspect the air ducts and make sure that there are no air leaks or discrepancies.
- Check the duct grilles and registers for proper air flow.
- If the oil furnace you are inspecting has a belt-driven blower motor, check the blower belt wear, tension, and adjust as needed.
- Check bearings and lubricate blower motor if needed.
- Check blower motor speed and adjust if necessary.
- Check voltages to the unit and check the voltage in the control box.
- Check the amperage of the blower motor and compare reading against the data information plate that is attached to the unit.
- Check the flue for rust and corrosion and advise of discrepancies.
- Check fuel lines and storage tanks for leaks.
- Check the heat exchanger and combustion chamber for cracks, corrosion, or soot buildup and advise of discrepancies.
- Check to see if flame ignition is instantaneous or delayed and advise of discrepancies.
- Check the barometric draft controls.
- Check, clean, and adjust burner if needed.
- Check oil pump operating pressure, oil filter, and nozzle.
- Check oil pump assembly.
- Check air tube and retention head.
- Check gap between electrodes.
- Check electronic ignition system.
- Check all wiring and connections to controls and electrical connections.

- Check transformer voltage.
- Check the CAD cell control.
- Check the stack control.
- Check the high-limit shut-off temperature.
- Check the CO_2 monitor and the smoke alarm for proper operation; replace batteries yearly.

Water Source Air Conditioning and Heating

- Check and adjust the thermostat.
- Check to make sure indoor and outdoor unit turn on.
- Check sequence of operation.
- Replace air filter or clean reusable type filter.
- Check and inspect the ducts to make sure that there are no air leaks or discrepancies.
- Check the duct grilles and registers for proper air flow.
- Check the evaporator coil and advise if dirty or if it needs cleaning.
- Check metering device and coil temperatures.
- If the unit you are inspecting has a belt-driven blower motor, check the blower belt wear, tension, and adjust as needed.
- Check bearings and lubricate blower motor if needed.
- Check the condensate drain system and secondary drain pan then advise of any discrepancies.
- Check condensate drain pan float switches for proper operation and installation.
- Check the condenser coil to determine if it needs cleaning.
- Check water source to air conditioning equipment.
- Check water source for cleanliness and proper operation.
- Inspect all wiring and connections to controls and electrical connections.
- Check voltage and amperage draw on all motors with a clamp-on multimeter.
- Check compressor contactor, capacitor, and any other controls in the condenser unit.
- Check the compressor and the amperage draw and compare the reading against the data information plate that is attached to the unit.
- Check reversing valve for proper operation.
- If the unit you are inspecting has a start capacitor and potential relay, check it.
- If the unit you are inspecting has a pressure switch, check the cut-out settings.
- Install refrigerant gauges and check operating pressures, superheat, and subcooling.
- Check refrigerant levels and advise the owner if adjustment or repairs are necessary.
- Check the conditions of the entire air conditioning system and advise the owner of any discrepancies.

Air Duct Distribution Preventative Maintenance

Residential forced air conditioning and/or residential forced heating systems efficiency is affected by the following:

- Air duct leakage
- Air conditioning and/or heating system airflow
- Blower operation
- Air ducts balanced between the supply air and the return air
- Condition of the air duct system and insulation

If the air ducts are not inspected, adjusted, repaired, or replaced first, the air conditioner refrigerant charge cannot be checked properly or recharged correctly with refrigerant. You will have poor superheat or poor subcooling temperatures. As for the heating systems, if the air ducts are not inspected, adjusted, repaired, or replaced, the heating system might not function properly or it might overheat and the system will shut down. HVAC contractors and service technicians servicing HVAC equipment must also determine if the air duct distribution system is adequate for the HVAC equipment. Undersized ducts will decrease efficiency and increase energy costs.

Inspecting Air Ducts

Before starting your quest through the attic or underneath a home, you will need to prepare the duct system. You must bring a trouble light and a flashlight, to guide you through the darkness; bring a mirror too, this will help you see and locate leaks under the ducts. Put on safety glasses and wear a dust mask, this will prevent you from breathing the dust in an attic when you begin your inspection. One of the simplest ways of finding air duct leaks is feeling with your hand for air leaking from the supply ducts, while the ducts are pressurized by running the air-handler blower fan. Closing the dampers on supply registers temporarily or by blocking the register with plastic, or with any object that won't be blown off by the registers airflow. This will increase the duct pressure within the duct and makes looking for air leaks easier. When inspecting residential air ducts look for the obvious (Figures 10-1a to 10-1k).

Another way to locate air leaks is to insert a trouble light with a 100-W bulb inside the duct through a register. Look for light emanating from the exterior of the duct joints and seams. Some technicians have in their arsenal of tools a video inspection camera, this is used to inspect the duct system from the inside (Figure 10-2).

So, when was the last time your air ducts were inspected? There are thousands of duct installations in residential homes that are installed correctly and according to code. But, if you are a duct installer or service technician in need of updating your duct skills, or you are an entry-level HVAC technician or installer, the following companies offer training and literature on proper construction, installation, and repairs of forced air ducts:

- North American Insulation Manufacturers Association (NAIMA) is the association for manufacturers of fiber glass, rock wool, and slag wool insulation products. NAIMA has a great library that you can access. Their Web site is *http://www.naima.org*.

Figure 10-1 (*a*) Air supply duct improper repairs made; (*b*) air supply duct has been patched with tape—this duct should have been replaced; (*c*) the return duct supply has improper repairs made; mold is starting to appear on the duct—this return duct should be replaced; (*d*) the plenum is separating from the air handler and leaking cold air; this duct needs to be replaced and installed correctly; A 2-in high metal flange needs to be installed on the air handler and the new plenum attached correctly with 2-in square metal plates and screws on all sides and taped properly; (*e*) tape separated from plenum and leaking cold air; this unit also needs 2-in metal flange installed and secured properly, then taped; (*f*) look closely, electrical wires should not be protruding from the inside of the air ducts; this is illegal to do and is a fire hazard; (*g*) when inspecting the air ducts, look for wet spots on the drywall ceiling; this is a sign that the air duct system is leaking cold air; the hot air in the attic, mixed with the cold air from the ducts, causes wet spots to appear along with mold; (*h*) the view is looking into the plenum, you can see the duct has separated from its outer duct liner and duct section is about to fall on the electric heaters; (*i*) this duct elbow is poorly constructed and it is splitting at the seams; this elbow needs to be replaced and installed properly; (*j*) air ducts are separating and leaking cold air—repairs are needed; and (*k*) air duct improperly constructed—leaking cold air and mold is beginning to grow. (*continued*)

Figure 10-1 (*a*) Air supply duct improper repairs made; (*b*) air supply duct has been patched with tape—this duct should have been replaced; (*c*) the return duct supply has improper repairs made; mold is starting to appear on the duct—this return duct should be replaced; (*d*) the plenum is separating from the air handler and leaking cold air; this duct needs to be replaced and installed correctly; A 2-in high metal flange needs to be installed on the air handler and the new plenum attached correctly with 2-in square metal plates and screws on all sides and taped properly; (*e*) tape separated from plenum and leaking cold air; this unit also needs 2-in metal flange installed and secured properly, then taped; (*f*) look closely, electrical wires should not be protruding from the inside of the air ducts; this is illegal to do and is a fire hazard; (*g*) when inspecting the air ducts, look for wet spots on the drywall ceiling; this is a sign that the air duct system is leaking cold air; the hot air in the attic, mixed with the cold air from the ducts, causes wet spots to appear along with mold; (*h*) the view is looking into the plenum, you can see the duct has separated from its outer duct liner and duct section is about to fall on the electric heaters; (*i*) this duct elbow is poorly constructed and it is splitting at the seams; this elbow needs to be replaced and installed properly; (*j*) air ducts are separating and leaking cold air—repairs are needed; and (*k*) air duct improperly constructed—leaking cold air and mold is beginning to grow.

(a)

(b)

(c)

FIGURE 10-2   (a) A view from inside of the flexible duct with a video camera—you can determine if the duct is coming apart or blocked with debris; (b) an internal view of a fiber glass duct—you can see the duct is intact, but is in need of cleaning; and (c) in this picture, you can see the duct and turning vanes—this inspection shows dirty duct and dirt accumulated on the turning vanes.

- HVAC Fiber Glass Duct Systems Trainers, fiber glass duct fabrication and installation training. Their website is *http://www.ductinstruct.com.*

- The Air Diffusion Council represents North America's manufacturers of flexible air ducts and air connectors. They also have online training for anyone who wants to or needs to learn about flexible ducts. Their Web site is *http://www.flexibleduct.org.*

- Sheet Metal and Air conditioning Contractors' National Association (SMACNA). They have an educational program and a publication store to order training manuals. Their Web site is *https://www.smacna.org.*

Performing Preventive Maintenance on a Residential Air Conditioning System

First, check the thermostat (Figure 10-3). Are the settings set in the correct position? On digital thermostats, some models are battery-operated; check the batteries (Figure 10-4). Sometimes the batteries may be leaking or the batteries are weak.

FIGURE 10-3
Check thermostat
settings and
programming.

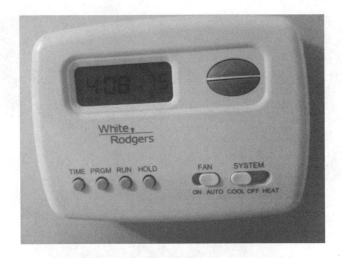

FIGURE 10-4
Inspect and test
batteries in
thermostat. Some
model thermostats do
not require batteries.
Check with the
thermostat literature.

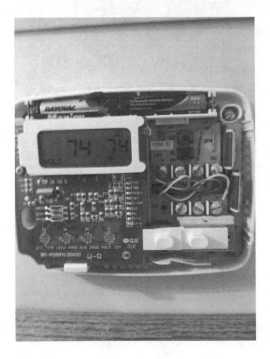

I recommend that you replace the batteries at least once a year. Locate the air handler (Figure 10-5); is it running? While you are at the air handler, check the filter, replace if necessary (Figure 10-6). Place your hand in front of a register (Figure 10-7); do you feel the air blowing out? On cooling days, does the air feel cool? On heating days, does the air feel warm? Next, locate the condenser (Figure 10-8); is it running? At this point in time the air

FIGURE 10-5   Check air handler; make sure it's running.

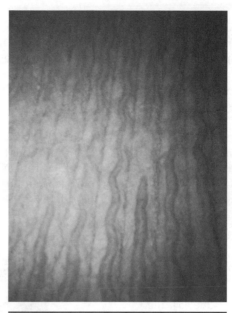

FIGURE 10-6   Check and replace the filter. Replace the filter with a high MERV rating.

FIGURE 10-7   Check all grilles and registers for air flow and if they are adjusted properly.

FIGURE 10-8   The condenser unit should be on when the thermostat is calling for cooling. If the unit is a heat pump, it should be on when the thermostat is calling for heat.

conditioning equipment is on and running. If the unit you are checking is not running, perform diagnostics and correct the fault with the unit. The next step is to check the grilles and registers make sure they are all open.

Preventive Maintenance on the Air Handler

Before accessing the air handler, turn off the electricity to the unit. To gain access to the evaporator coil, blower assembly, controls panel, electric heaters, condensate drain, remove the front access panels from the air handler or disassemble the unit per the instructions given in the service literature. Next, inspect the evaporator coil for damage, rust, refrigerant oil leaking, or dirt accumulation. A flashlight or drop light will be helpful in the task. Is it clean or dirty (Figures 10-9, 10-10, and 10-11)? If the evaporator coil needs

FIGURE 10-9
Evaporator: A-coil view of the backside of the coil. Air has passed through the coil. The appearance indicates a clean coil. Remember to follow the air-flow pattern; this will indicate which way the air flows through the coil.

Direction of air flow

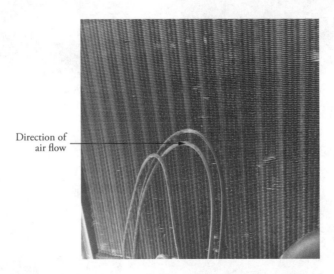

Direction of air flow

(a)

Clean coil with a UV lamp installed

(b)

FIGURE 10-10 (a) Evaporator: (a) coil view of the front side of the coil—air will pass through the coil; and (b) evaporator coil clean—this unit has a UV lamp installed to keep slime and algae off the coil.

Direction of
air flow →

FIGURE 10-11 Evaporator M-coil view of the backside of the coil. This dirty coil has blocked the air flow completely. When the air conditioning unit is running with a dirty coil, the evaporator coil will begin to freeze up. The end result, no cooling.

cleaning, use a vacuum and a brush attachment to clean the coil surface and fins. When brushing the fins, always go in the direction of the fins, as not to bend the fins. For stubborn dirt removal, oil or grease, use a tank-type sprayer to spray the coil with a mild soap solution and water or approved coil cleaning solution.

NOTE *Do not use an alkaline solution with a pH of 9.5 or greater. It will damage the coil.*

Next, rinse the coil thoroughly with clear water. Be careful not to spray the cleaning chemical or water on the motor, electrical panel, wiring, insulation, or on the air filter. Also, clean up any spillage in the surrounding area. If any fins are bent, use a fin comb to straighten the coil fins.

Gain access to the condensate drain pan, and make sure it's clean and the condensate water is flowing (Figures 10-12 to 10-14). Following are the ways to clean out the condensate drain pan:

1. Use a wet vacuum to suck up the condensate water and debris.

2. Remove all algae by washing the condensate pan with a solution of household bleach and/or an approved commercial chemical solution.

3. If the air handler is located in the finished ceiling or living space, make sure that the secondary condensate drain pan and piping are not clogged (Figure 10-13). When cleaning the secondary condensate drain pan and piping, clean in the same manner as cleaning the primary condensate drain pan and piping.

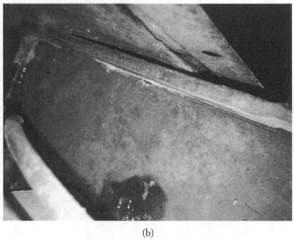

(a)   (b)

FIGURE 10-12   (*a*) Check the condensate drain pan; clean as needed; (*b*) make sure it's clean and the condensate water is flowing down the drain line.

FIGURE 10-13
View of a horizontal
air handler. Drain pan
overflowing and
leaking water through
ceiling.

Secondary drain
port and pan

FIGURE 10-14
Condensate water overflowed the drain pan and is accumulating in the other section of the air handler. Condensate water will have to be removed with a wet and dry vacuum and dried out to prevent mold from growing.

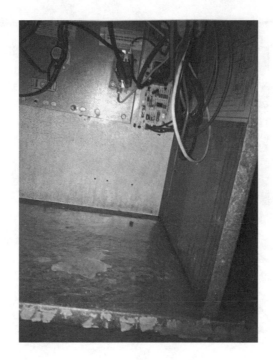

4. Clear out the blockage in the condensate drain line, with a wet and dry vacuum cleaner or use water pressure from a hose to free the blockage in the condensate line. Sometime you may need to use a plumber's snake to clear the line.

If the air handler has a condensate pump (Figure 10-15), it too will need to be inspected and cleaned out. Make sure the condensate pump is unplugged or turn off the circuit breaker before attempting servicing. Next, remove the condensate pump from the unit and drain the water from the reservoir. The condensate pump pan can be separated from the reservoir for cleaning. Thoroughly clean the reservoir of any grime and dirt. Clean the float and make sure it moves freely. If you are uncertain on how to service the condensate pump, follow the instructions in the service literature from the condensate pump manufacturer.

When you have completed the maintenance on the condensate pan(s), check with local codes and laws or equipment manufacturers, if you can place approved chemicals or tablets in the condensate pan to help prevent the buildup of slime or algae.

For added protection from condensate water leaks, install an emergency condensate safety shut-off switch if the unit you are servicing does not have one installed. This device will shut off the air conditioning unit in the event of the condensate pan filling up and not draining the condensate water.

With the circuit breaker off to the air handler, inspect the electric controls and wiring (Figure 10-16). Make sure all connections are tight and not loose on the controls. Repair or replace any burnt wiring or connectors. Next, inspect the electric heaters; Make

FIGURE 10-15
If the air handler has a condensate pump attached, it too must be inspected and cleaned.

Circuit breaker

FIGURE 10-16 Turn off the circuit breaker first. Inspect the controls and wiring in the control box, make repairs if necessary.

FIGURE 10-17   Inspect the electric heating elements for debris and breakage in the heating element wires.

sure that there is no debris on the heater elements (Figure 10-17), or broken heater wires. Electric heater(s) need little maintenance. When the heaters are on they will burn off any dust from the elements. This is normal.

Inspect the fan blower assembly and motor (Figures 10-18 and 10-19). If there is dust on the motor, you will have to vacuum the dust off the motor and clean the dust off the blower wheel. If dust is allowed to accumulate on the motor, it over heat and the life expectancy of the motor will be shortened. Vacuum the dust from the inside cabinet of the air handler. This should improve air quality to the conditioned space.

FIGURE 10-18   Inspect the blower wheel, and clean any dirt from the blower vanes.

FIGURE 10-19   Inspect and clean the dust from blower motor. This will increase the motor life expectancy. Lubricate the motor as needed. Some motors are permanently lubricated motors.

Some air handlers or package units have a UVC lamp installed in it (Figure 10-20). The UVC lamp reduces microbial growth within the air handler or package unit, allowing the air conditioning equipment to run more efficiently, improving indoor air quality within the occupied space. The UVC lamp prevents mold, mildew, bacteria, and viruses from collecting inside the air handler and attached plenum area. This lamp must be inspected

FIGURE 10-20
Some air handlers have UVC lamps installed. Check UVC lamp operation. Replace UVC bulbs when necessary or recommended by the UVC manufacturer. See UVC lamp service literature for proper instructions.

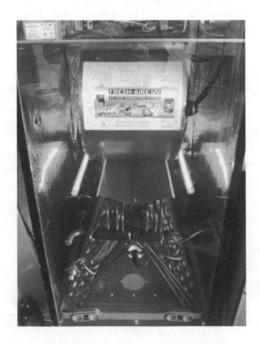

every time preventive maintenance is done on the air conditioning equipment. With the electricity turn off to the UVC lamp, wipe the dust from the bulb. Depending on which type of UVC lamp installed, you might have to replace the UVC bulb every 1 to 2 years. Check the service literature that came with the UVC lamp assembly.

SAFETY NOTE *Human exposure to UVC light may result in temporary eye (cornea) damage and skin (sunburn) damage. While these effects are mostly temporary, they can still be very painful. Service technicians servicing air conditioning equipment that has UVC lamps installed in it should be aware of hazards and take proper measures for self-protection. The first step when servicing an air conditioning unit, turn off the electricity to the air conditioning unit or UVC lamp assembly. If for any reasons the technician needs to directly view the lamps, follow these guidelines:*

- When temporarily viewing lamps from 4 ft or more, you will need to wear eye protection with polycarbonate lens (rated for UV) and the lenses must wrap around the face to fully protect the eyes (goggle-style eyewear).

- The most critical, when directly exposed to UVC less than 4 ft, it is necessary to wear full face and skin protection. This viewing is not suggested unless the service technician has been properly trained.

Preventive Maintenance on the Condenser

Locate the condensing unit (Figures 10-8 and 10-21). Turn off the electricity to the unit. Next, gain access to the controls (Figure 10-22). Inspect the controls and wiring. Make sure the wires are in good condition with no burn marks on them. Make sure all wire connections are tight. If you diagnose defective components, replace as necessary.

FIGURE 10-21
Locate the correct condensing unit. Turn off the electricity to the unit.

FIGURE 10-22 Inspect the control box. Make sure all wiring connections are secure and tight. Inspect all the components; make sure they are in good condition.

Gain access to the condenser coils by disassembling the unit per the instructions given in the installations instructions or service manual. Inspect the compressor, condenser coil, and filter drier(s) (if the condensing unit has one installed on the suction line or liquid line) for rust or refrigerant and oil leakage.

Inspect the fan motor and fan blade. Make sure the fan blade has no cracks on the fan paddles. If the motor has oil port, lubricate the motor. Next, inspect the condenser coil and clean if necessary. If the fins are bent, straighten them out with a fin comb.

Clean the condenser coil in the same way as previously described for the evaporator coil cleaning. Spray the cleaner solution from the inside to the outside of the coil. Make sure to flush all dirt and debris from the base of the condenser unit. Last, reassemble the condenser unit in the reverse order of disassembly.

Make sure no shrubs or plants are blocking the air flow of the condenser (Figure 10-8). If they are, advise the owner to have the shrubs and plants cut back at least 24 in from the condensing unit.

When performing preventive maintenance on packaged air conditioning units (Figure 10-23) or heat pump condensing systems, perform the maintenance tasks the same way as previously described in this chapter. On heat pump models, make sure the reversing valve is operating according to manufacturer's specifications listed in the service literature (Figure 10-24).

Checking the Cooling System for Proper Refrigerant Charge

Only an EPA-certified technician can check the refrigerant charge in the air conditioning unit. For a non-thermostatic expansion valve (TXV) systems, the technician will have to check the refrigerant charge for the proper superheat according to the

FIGURE 10-23   A package unit has both the evaporator and condenser coils, compressor, condenser fan motor and blade, blower fan assembly, and controls all packaged in one unit.

FIGURE 10-24
A heat pump refrigerant reversing valve.

manufacturer's specifications as listed in the service literature. For TXV systems, the technician will have to check the refrigerant charge for the proper subcooling according to the manufacturer's specifications as listed in the service literature.

Checking for Proper Airflow

For proper cooling in an occupied space, the indoor blower should be moving from 400 to 450 cubic feet per minute (CFM) of air across the evaporator coil for each ton (12,000 BTU) of cooling capacity. If there is too much or too little air across the evaporator coil, the comfort level in the building will be affected and the refrigerant charge level will be affected also. To properly check the CFM of the unit, consult the manufacturer's service literature for the proper procedure for measuring the amount of airflow required.

Preventive Maintenance on a Gas Furnace

Standard minimum maintenance on the gas furnace should include the maintenance on the air filter$_1$ (clean or replace), blower fan motor assembly (remove dust by vacuuming) (Figures 10-18 and 10-19), and the electrical components (run diagnostics when necessary) (Figures 10-25 and 10-26). Check the conventional thermostat's heat anticipator setting. The thermostat's heat anticipator setting should match the measured current (amperage) in the 24-V control circuit. For programmable thermostat, you must refer to the manufacturer's instructions about how to control the cycle length. These instructions may be printed on the inside of the thermostat. If the thermostat model you are servicing has batteries, replace

FIGURE 10-25
Inspect the wiring and connectors.

FIGURE 10-26
Make sure the inducer
motor assembly is
free of dust.

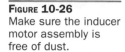

them yearly or when they become weak (Figure 10-4). The gas furnace should be inspected by a qualified service company or service technician at least once a year. The inspection and maintenance should be performed just before the heating season. This will ensure that all the components in the furnace are in proper working order and that the heating system functions are working appropriately. The purpose of the maintenance procedures and tests is to reduce carbon monoxide (CO) emissions, stabilize the burner flame, and verify the operation of the safety controls.

The service technician should pay particular attention to the following items. Make repairs if needed, perform maintenance, or replace defective components as necessary:

- *Flue pipe system.* Check for obstructions and/or leaks in the piping system (Figures 10-27 and 10-28). Check the outside termination and the connections at and internal to the gas furnace. Check the venting pipe system for the proper diameter, pitch, length including elbows used, and type of vent pipe used (Figures 10-29 to 10-31). Check with the manufacturer's specifications in the installation guide.

- *Combustion air intake pipe system (if applicable).* Check for obstructions and/or leaks in the piping system. Check the outside termination and the connections at and internal to the gas furnace. Check the venting pipe system for the proper diameter, pitch, length including elbows used, and type of vent pipe used. Check with the manufacturer's specifications in the installation guide.

- *Heat exchanger.* Check for corrosion and/or buildup within the heat exchangers passageways (Figure 10-32). Check the heat exchanger for leaks and cracks. Clean the heat exchanger, if there are signs of soot around the burner compartment. On natural draft furnaces, clean any accumulation of dust, or light dirt. However, the cleaning of the heat exchangers in condensing furnaces and induced draft furnaces is not routinely performed. The cleaning of the heat exchanger in these types of

FIGURE 10-27 Inspect flue pipe and test the gas pressure with a manometer.

FIGURE 10-28 Check the vent terminations.

FIGURE 10-29 Check vent pipe connections.

FIGURE 10-30 Check elbows, pitch, and the length of vent pipe.

FIGURE 10-31
Check on the type of vent pipe used. For example, you cannot use type "B" vent pipe on a condensing furnace. It will rust out in less than 6 weeks.

FIGURE 10-32 Inspect heat exchanger for leaks, cracks, rust, or corrosion.

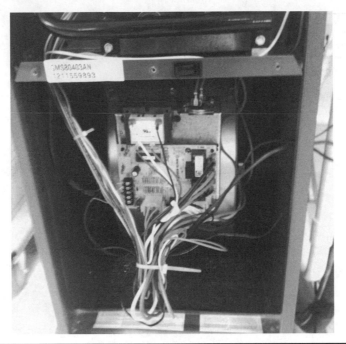

Figure 10-33   Inspect wiring and connections. Look for melted wiring.

furnaces is a complex procedure. This is because the disassembly and reassembly might cause damage to the gaskets and seals that are disturbed in the process. Always follow the procedures for cleaning in the service literature.

- *Burners.* Look for soot, melted wire insulation (Figure 10-33), and rust in the burner (Figure 10-34) and manifold area outside the burner box. These signs indicate flame roll-out, combustion gas spillage, and carbon monoxide (CO) production. Inspect the burners for dust, debris, misalignment, flame impingement, and other flame interference problems. Clean, vacuum, and adjust as needed. Check for proper ignition, burner flame size, and flame sense (Figures 10-35 to 10-38). On standing pilot models, determine that the pilot is burning and that main burner ignition is satisfactory. Check in the service literature for the proper adjustment as needed.

- *Drainage system.* Check for blockage and/or leakage in the condensate drainage system. The condensate drainage system for a condensing furnace should always be cleaned according to the instructions given in the service literature. Inspect the collector box tube and inducer housing drain tube to make sure that they are not pinched, kinked, cracked, brittle, or otherwise damaged. Repair or replace any restricted or damaged tubes (Figures 10-34 and 10-39). This will ensure proper condensate drainage. By performing the maintenance on the drainage system; it will prevent condensate buildup, and water damage or corrosion in the furnace. If the condensate drainage maintenance is not performed, it is a good possibility that the furnace will fail to operate.

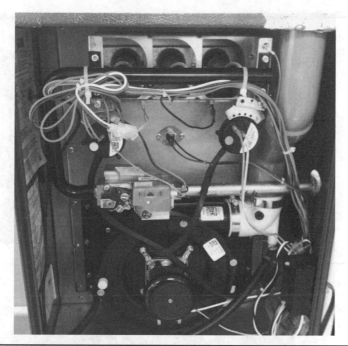

FIGURE 10-34   Look for rust in the burner area.

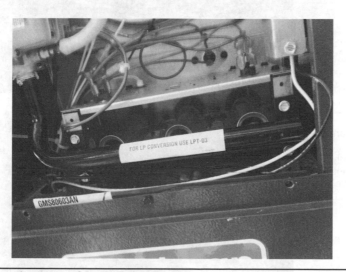

FIGURE 10-35   Three burner gas furnace. Inspect the burners.

FIGURE 10-36
Only the middle burner
is burning. Check gas
pressures.

FIGURE 10-37
The two outer burners
are burning. Check for
blockage, check gas
pressures.

FIGURE 10-38
All three burners are
burning. The left
burner is improperly
burning. The service
technician will have to
run gas diagnostics.

Figure 10-39   Inspect condensate drainage system for blockage and leaks.

- *Wiring.* Check the electrical connections for tightness. Check for corrosion on the terminal ends and connectors. Inspect all wires internal to the furnace (Figures 10-33 and 10-34).

- *Other components located in the furnace.* Check the pilot safety and test it for complete gas valve shutoff. Next, test to ensure that the high-limit control extinguishes the burner before the furnace temperature rises to 200°F. Measure the gas input pressure at the gas valve (Figure 10-27); and observe the flame characteristics. Is the flame producing soot, carbon monoxide (CO), or other combustion problems? Make necessary adjustments and/repairs as needed. Vacuum off any dust from the inducer fan assembly (Figure 10-26). Any dust accumulation on this motor will shorten the life expectancy of the motor (Figures 10-18 and 10-19).

Performing Maintenance on a Pilot/Direct Ignitor System

Clean the components of a standing or spark re-ignition pilot assembly of any dirt, soot, scale, or carbon by using a soft bristle brush. In a spark re-ignition pilot, inspect the spark electrode for any cracks and make sure the spark electrodes are positioned according to the specifications in the service literature. Next, check the high-voltage lead to the electrode; inspect for good contact and no cracks. Replace if necessary.

Hot-surface ignitors require no maintenance. However, they can develop tiny cracks that can cause a failure. When handling this type of ignitor, care must be taken as not to damage it when it is removed. This type of ignitor is very fragile. If you suspect a crack in the ignitor, use a multimeter set on ohms, and test the resistance. You should read between 45 and 90 Ω.

Resistance readings on the multimeter of over 105 Ω and higher are a definite indication of a crack. If have a cracked ignitor, replace with a duplicate of the original.

Combustion Air System and Operating Checks and Adjustments

In tightly constructed buildings, or a confined space, make sure that the openings, grilles, ducts, and the like are open and not blocked or restricted by any shrubby, boxes, or stored materials, or other obstructions that will prevent fresh air from entering the combustion area, causing improper combustion and possibly become a fire hazard. For gas furnace models that require indoor air for combustion, make sure that all hazardous materials and chemicals are not stored in the same area as the furnace. Vapors from these items can contaminate the combustion air causing premature rusting of the heat exchanger and possible premature furnace vent failure causing a fire hazard and bodily harm. If this condition exists, advise the owners to store the materials away from the furnace.

After all inspections and maintenance procedures are completed, the gas furnace is ready for operational checks and adjustments including testing the safety controls for proper operation. The operating checks and adjustments include:

- Sequence of operation
- Pilot checks
- Burner checks
- Gas pressure check
- Temperature rise check
- Thermostat heater check
- Safety control checks

It is recommended for the proper sequence of operation that the technician must consult the service literature for the model that is being serviced. All other checks must be checked according to the furnace manufacturer's recommendations in the service literature. Failure to follow the manufacturer's recommendations will cause equipment failure, bodily harm, or death.

Checking the Sequence of Operation on a Gas Furnace

Turn on the gas furnace according to the manufacturer's instructions and run the furnace through two complete heating cycles. The furnace operating sequence will vary depending on which type of ignition system is used in the furnace. During the operation of the furnace, check the condition of the following:

- Pilot and/or burner flames
- The time it takes to ignite the pilot and/or burners
- The operation of inducer and/or blower motors
- Condensate flow

- Check the furnace sequence of operations with the service literature sequence of operation to compare.

Checking the Pilot on a Gas Furnace

While the furnace is operating, check that the standing or re-ignition pilot flame has the correct color and size. The flame should extend above the burner carryover port so that it will properly ignite the burners. The pilot flame should be blue in color with some yellow at the tip. Check the height of the pilot flame; it should be 3/8 to ½ in long to impinge on the thermocouple or flame sensing electrode. If the pilot needs adjusting, adjust the pilot pressure adjustment screw located on the gas valve to make the pilot smaller or larger. Check with the service literature to see if the furnace you are servicing has this feature to adjust the pilot flame. Some models do not. If the pilot flame has a yellow or orange color, this indicates that dirt or lint has entered the pilot orifice. When this happens, the dirt or lint will change the air-to-gas ratio. To clean a dirty orifice do the following:

- Turn off the gas supply to the furnace.
- Turn off the electricity to the furnace.
- To gain access to the pilot assembly, follow the manufacturer's instructions in the service literature to gain access.
- Disconnect the gas line and thermocouple (if applicable) from the pilot assembly.
- Remove the pilot orifice from the pilot assembly.
- Clean the pilot orifice with a wire, do not damage the orifice opening. If need be, blow air through the orifice. If the orifice is damaged, replace it with a duplicate of the original.
- After cleaning the orifice, reassemble the pilot assembly and reinstall it back into the furnace. Make sure the installation of the pilot assembly and positioning complies with the specifications provided in the service literature.
- Turn on the gas supply, and then turn on the electricity to the furnace.
- Relight the pilot and/or operate the furnace as necessary to recheck the pilot flame.

Checking the Burner

When the furnace is first turned on and the burners are ignited, watch the flames as the furnace blower turns on. If the flames are distorted or flicker, it is an indication that the heat exchanger is cracked. Shut off the gas immediately, and then perform a safety check on the heat exchanger.

If there is no cracks in the heat exchanger, run the furnace for 5 minutes, making sure the burners are hot enough, check the condition of the main burner flames. The flames should be quiet and stable and be a clear blue in color, almost transparent, with a well-defined inner cone.

To adjust the burners, consult the service literature for the model you are servicing. Note that too much primary air can produce a well-defined flame that has a tendency to float off the burner ports.

Checking the Vent Damper and Vent Pipe System

On furnaces with dampers, make sure they are open when the burners are on and closed when the burners turns off. On natural draft furnaces, after the furnace has been running for 5 minutes, use a match flame or candle moving it around the perimeter of the draft inverter, to check and see if the air is drawing in the flame and smoke. This will indicate a properly operating vent system.

Checking the Temperature Rise in a Gas Furnace

The difference between the air temperature entering the furnace and leaving the furnace is called temperature rise. The amount of temperature rise gives an indication of whether adequate air is flowing across the heat exchanger. To locate the temperature rise on the model furnace you are servicing look on the information data tag located on the furnace. In condensing furnaces, if too much air passes over the heat exchanger, condensing can take place in the primary heat exchanger, causing corrosion and failure. In natural draft and induced draft furnaces, too much air can cause condensation in the heat exchanger and vent pipe system. If too little air passes over the heat exchanger, the outcome is overheating and failure. The procedure for measuring temperature rise is listed in the service literature for the furnace.

Checking the Thermostat Heat Anticipator

For most thermostats, the heat anticipator that is located in the thermostat will have to be checked to make sure the furnace will not short-cycle. Short-cycling prevents the furnace from coming up to operating temperature, causing condensation in the vent pipe and heat exchanger. Short cycles or excessive long cycles contribute to poor indoor comfort for the occupants. There are many models and types of thermostats on the market today and it will be almost impossible to list all of them in this chapter. Check with the manufacturer for the thermostat model you are servicing or the service literature that comes with the thermostat to make the correct adjustments to the heat anticipator. Regardless of which model thermostat is installed, the furnace must make two complete burner cycles to ensure proper furnace operation.

Checking the Safety Controls in a Gas Furnace

The following safety components must be checked for the proper operation as listed in the service literature:

- *High temperature limit switch.* If the blower motor fails or if there is a restriction in the air supply, the gas furnace will overheat. To check this safety switch, the technician will slowly close the air supply while the burners are burning and the blower is on. The limit switch should turn off the burners. If the burners are off, this will confirm the operation of the limit switch. Then, unblock the air supply to permit normal operation of the burners. If the high limit switch fails to shut off the burners within a reasonable time, this could be an indication of a bad limit switch.

- *Induced-draft pressure-sensing switch.* Before turning on the gas furnace to start the heating cycle, disconnect the sensing tube from the related switch, this should prevent the burners from burning when the furnace is calling for heat. If the burners ignite, then the switch will need to be replaced. Reinstall the sensing tube to restore operation.

- *Standing pilot shutoff.* Test the pilot shutoff with the furnace operating. Turn the electricity off to the furnace. The burners should turn off. Blow out the pilot light. Usually within three minutes, the solenoid drops out, shutting off the gas flow. If the gas flow continues, the gas valve will need to be replaced. Now, restore the electricity to the furnace, there should not be any gas flowing to the burners. Relight the pilot to restore normal operation to the furnace.

Preventive Maintenance on an Oil Furnace

Oil furnaces should be checked and maintenance performed before the heating season begins. I recommend that a certified technician perform the preventive maintenance on the oil furnace. The following maintenance should be performed:

- Clean and/or replace filters monthly as required.
- Clean, inspect, and lubricate the blower motor assembly.
- Check the condition of the electrical wiring and tightness of terminals and connections.
- Clean and inspect the heat exchanger and combustion chamber.
- Clean and inspect the flue pipe system, chimney/vent, and draft regulator.
- Check the oil tank and the connecting piping for leaks.
- Clean and inspect oil burner assembly; lubricate burner motor, replace the oil nozzle, adjust the electrodes, and clean the oil pump strainer (if used).
- Replace the oil supply line filter.
- After the maintenance has been performed on the oil burner assembly and filter, on a single-pipe system, bleed the system of air.
- Check oil pump pressures.
- Check combustion air ducts, grilles, and the like.

Operational checks and adjustments

- Check the sequence of operation against the service literature for correct operation of the furnace.
- Check the flame characteristics.
- Perform combustion efficiency checks/tests; overfire draft and flue draft, net stack temperature, carbon dioxide (CO) percentage, smoke spot test, determine combustion efficiency.
- Check temperature rise.
- Check the thermostat heat anticipator setting.
- Check safety controls (high-temperature limit switch, flame cutoff time, and the like).

Combustion System Maintenance

When performing maintenance on the combustion system, make sure the electricity is turned off to the furnace. Also, make sure the furnace is not hot when performing maintenance. Clean and inspect the combustion chamber and heat exchanger according to the information provided in the service literature. Following are some general guidelines for cleaning and inspecting the combustion chamber and heat exchanger.

SAFETY NOTE *You never want to operate an oil furnace with a cracked heat exchanger because the products of combustion will enter the residence and cause sickness and/or death of occupants. When inspecting the oil furnace, and you notice a cracked heat exchanger, replace the furnace.*

- Turn off the oil supply to the oil burner assembly.
- Turn off the electricity to the oil furnace.
- Inspect the combustion chamber (Figure 10-40).
- Inspect and clean the oil furnace blower assembly (Figure 10-41).
- Follow the instructions for disassembly and excess of the heat exchanger in the service literature.
- Clean the heat exchanger with a long flexible brush. Work the brush end into the opening and move it in and out to clean the heat exchanger. Use a vacuum cleaner to vacuum up the loosened soot and debris out of the heat exchanger opening.

FIGURE 10-40   Clean and inspect combustion chamber and clean and inspect the heat exchanger.

FIGURE 10-41
Inspect and clean the oil furnace blower motor assembly.

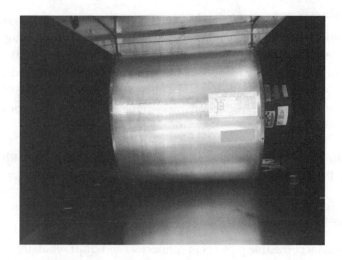

- Use a flashlight and mirror to inspect the heat exchanger condition. Make sure that there are no cracks in the heat exchanger.

- Disconnect the oil supply line from the oil burner, draining any oil remaining in the pump, into a container.

- Remove the bolts securing the oil burner to the combustion chamber (Figure 10-42); carefully remove the oil burner from the furnace.

FIGURE 10-42
Inspect, clean, and adjust the oil burner assembly.

- Vacuum the accumulated soot from inside the combustion chamber through the opening vacated by the oil burner. Be careful when vacuuming the soot, as not to damage the refractory material within the combustion chamber.

- Use a flashlight and inspection mirror to inspect the combustion chamber for any cracks, carbon deposits, loose insulation, or an accumulation of unburned oil.

- After cleaning and inspection, reassemble the furnace in accordance with the service literature.

Performing Maintenance and Inspection of the Flue Pipe and Vent System, and Draft Regulator

Clean and inspect the mechanical condition of the vent system per the guidelines given below. Make repairs as necessary.

- Disconnect the flue pipe (vent connector) and draft regulator from the main vent system. If the vent system is connected to a chimney, vacuum the base of the chimney.

- Use a flashlight and inspection mirror to inspect the vent system and/or chimney for corrosion, cracks, or obstructions. If you find bird nests, dead animals, deteriorated pieces of brick or mortar, or heavy accumulations of soot; make necessary repairs and/or remove obstructions. For masonry chimneys, inspect the exterior of the chimney for loose bricks, missing bricks, and mortar. For chimneys with metal liners, open the cleanout door and inspect for signs of corrosion. Make repairs as necessary.

- Inspect the draft regulator and vent pipe for corrosion or other damage. Make sure the draft regulator damper gate swings freely. Remove any soot, scale, and/or dirt from the vent system pipe and draft regulator.

- Finally, reinstall the draft regulator to the vent pipe and reinstall the vent pipe according to the manufacturer's recommendations.

Oil Burner Maintenance

With the oil burner removed from the furnace (Figures 10-42 to 10-46); inspection, cleaning, lubrication, and adjustment can be performed. On some model furnaces, the oil burner maintenance can be performed with the oil burner in place. Always follow the manufacturer's recommendations for proper maintenance of the oil burner.

SAFETY NOTE   *Whenever working on the oil burner, avoid spilling fuel oil, it is a fire hazard and the oil leaves an odor.*

Follow the following guidelines when performing maintenance on the oil burner:

- Shut off the oil supply to the oil burner.
- Turn off the electricity to the furnace.

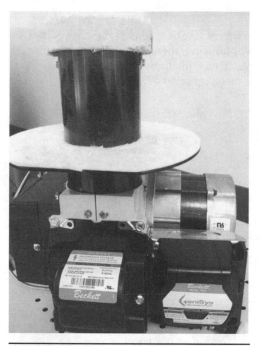

FIGURE 10-43   Rear view of the oil burner assembly removed from the furnace.

FIGURE 10-44   Side view showing the oil pump.

FIGURE 10-45   A view of oil nozzle and combustion head.

FIGURE 10-46   Side view of the oil pump fan blower assembly.

Burner air delivery component, inspect and clean

- To move the ignition transformer from its current location, loosen the screws that attach the ignition transformer to the blower housing, then move the transformer out of the way by tipping it up on its hinges. This will allow you access to the blower wheel and drawer assembly below (See Figure 10-47).

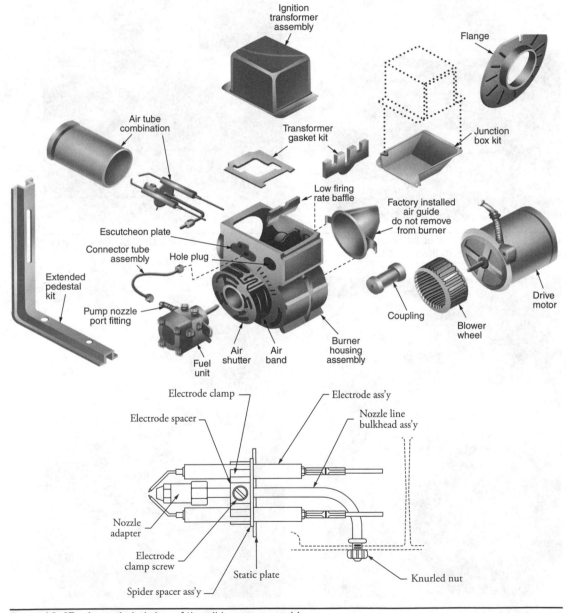

FIGURE 10-47   An exploded view of the oil burner assembly.

- Remove the nozzle oil line from the housing wall located next to the blower wheel. Gripping the nozzle oil line and the electrode assembly, pulling firmly toward] you, lift out the electrode assembly from the air tube.
- Clean the blower wheel with a soft brush and vacuum up the dust.
- Clean and vacuum soot from the burner assembly air tube.
- Clean and vacuum the air intake slots on the blower motor.

Performing maintenance on the burner draw assembly

- Replace the nozzle, never try to repair or clean the old nozzle. When replacing the nozzle, always replace the nozzle with a duplicate of the original.
- Clean any soot and carbon from the electrodes and insulators, and check for cracks in the ceramic of the insulators. If you notice any cracked or worn insulators, replace as necessary.
- Check the positioning of the electrodes—look up in the service literature for the oil burner for the correct positioning of the electrodes. Use an electrode setting gauge to make the adjustments.
- Reinstall the nozzle-electrode assembly in the reverse order of disassembly.
- Reconnect the pump oil line located outside the housing wall to the nozzle oil line inside the housing.
- Clean the ignition transformer terminals/contacts.
- Using a clean cloth, clean the CAD cell lens. Make sure the CAD cell is positioned correctly and secured in place.
- Reinstall the ignition transformer and fasten it down.

Burner oil pump maintenance

- If the model oil burner you are servicing is equipped with an oil strainer, gain access to the strainer and remove for cleaning or replacement. When removing the strainer, use some rags to catch any oil spillage.
- Remove and discard the existing gasket that seals the cover and the pump body.
- Clean the strainer in a suitable solvent. If the strainer mesh is worn or bent, replace it with a duplicate of the original.
- Reinstall the strainer in the oil pump. Install the new gasket on the pump flange, then reassemble and fasten the pump cover to the pump.

Oil burner testing   Combustion testing is the key to a technician understanding the current oil burner performance level and potential for improvement. The following tests should be performed by a certified service technician:

- Analyze the flue gas of oxygen (O_2). Reading should be between 4 and 9%.
- Analyze the stack temperatures. Reading should be 325° to 600°F.
- Analyze the carbon monoxide (CO) levels. Reading should be ≥100 ppm.
- Sample the flue gases between the draft regulator and the furnace.

- Sample undiluted flue gases with a smoke tester, following the smoke tester instructions. Compare the smoke smudge left by the gases on the filter paper with the manufacturer's smoke spot scale to determine smoke number. Reading should be ≤1 *or* ≤2.

- Measure flue draft between the furnace and draft control and over fire draft over the fire inside the firebox. Reading should be overfire −0.02 IWC (inch water column). Reading for flue draft should be −0.040 to −0.1 IWC.

- Measure the high-limit shut-off temperature and adjust or replace the control if the shut-off temperature is more than 200°F.

- Measure the oil pump pressure, and adjust to manufacturer's specifications. The pump pressure should read between 100 and 140 psig.

- Measure transformer voltage; adjust to manufacturer's specifications.

- Time the CAD cell control or stack control to verify that the burner shuts off, within 45 seconds or a time specified by the manufacturer, when the CAD cell is blocked from seeing the flame.

- Adjust the fan speed or increase ducted airflow to reduce high flue-gas temperatures if possible without reducing the flue gas temperature below 350°F.

Fuel Oil Supply System Maintenance

The oil filter should be changed at least once a year to prevent contamination and plugging of the oil burner nozzle or the oil pump. The oil filter is located between the oil pump and the oil storage tank. To change the oil filter, follow the following guidelines:

- Close the oil shutoff on the oil storage tank side of the oil filter, to isolate the oil filter from the fuel supply.

- Place a waste oil container under the oil filter to catch any oil spillage.

- Remove the filter bowl from the filter lid.

- Remove and discard the oil filter element and gasket from the filter housing.

- Clean the filter housing and remove the grit from within.

- Install the new filter element and gasket according to the manufacturer's specifications.

- Open the oil filter bleeder fitting, open the oil supply line valve; when oil comes out of the bleeder port, close the bleeder valve.

- Clean excess oil from the oil filter exterior.

Bleeding a one-pipe system   An oil furnace with a one-pipe system must be bled or purged to remove air that is trapped inside the system whenever the oil pump burner is serviced or when the oil filter is replaced. Also, when the storage tank is empty, the oil system must be bled or purged. A two-pipe system never needs to be bled or purged, since they are self-bleeding. Follow the guidelines when bleeding or purging a one-pipe system:

- Connect a ¼-in flexible, transparent tubing to the oil pump vent plug. Place the other end in a waste oil container.

- Turn on the electricity to the oil furnace and open the oil supply line shutoff valve.
- Set the room thermostat controls, set the temperature above the room temperature. Then, open the oil pump bled port valve; bleed the oil until you see clear oil coming out. Close the bleed valve, the burner should fire when the oil reaches the nozzle.
- Turn off the furnace, testing is now completed.

Operational Checks on Air Conditioning Equipment

After the cleaning and mechanical inspections are completed on the air conditioning and heating equipment, the operation of the air conditioning and heating equipment must be checked to determine its operation according to the manufacturer's specifications. The operating checks and adjustments include:

- Input voltage checks to the equipment
- Current checks on the equipment
- Cooling system checks on the equipment
- Heating system checks on the equipment's electric resistive heaters
- Heating system checks on the heat pump

Sequence of operation   Information on the correct sequence of operation for the model unit you are checking is found in the service literature.

Input voltages and current to the air conditioning and heating equipment

- Test the voltage input and the current to the equipment while it is operating with a clamp-on ammeter.
- Compare the readings against the name data tag on the unit or with the service literature.

Water Source Air Conditioning and Heating Preventive Maintenance

To perform the preventive maintenance on a water source air conditioning unit, follow the steps listed under "Water Source Air Conditioning and Heating" in this chapter. Instead of a fan removing the heat from the refrigerant in the condenser coil, water is circulated through a "tube in a tube" type of condenser coil to remove the heat from the refrigerant. The water source can come from a surface water source such as a lake, river, or stream. The water can also come from a groundwater source such as a well or aquifer. Depending on the region, surface water will have seasonal variations and can carry silt and debris that can cause a fouling if the water is not pre-filtered before entering the condenser coil. Groundwater does not have seasonal variations. But, depending on the geology of the region, the water source can have high levels of dissolved minerals that contribute to scale formation on the inside of the condenser coil. The scale formation on the inside of the condenser coil acts as an insulator which prevents heat transfer from the refrigerant to the water. Some areas use a water tower that reuses the water by recirculating it from the tower to the cooling equipment and back to the tower again. Whichever type of water source is used, in due time the condenser coil will have to be cleaned. To clean the condenser coil follow the instructions listed in the equipment manufacturer's service literature.

Endnotes

1. Improper filter maintenance is the most common cause of inadequate heating or cooling performance.

Residential Refrigeration Appliances

This chapter will only cover the basics needed to perform the preventive maintenance on residential refrigeration appliances. Each preventive maintenance procedure has been rated by a degree of complexity or simplicity. For the owner of the product, you will have to decide whether you can perform the preventive maintenance procedure or if you must call a professional to complete the preventive maintenance procedures on the product. Following are symbols that are used in this chapter to define who should perform the preventive maintenance on the residential refrigeration appliances.

The home owner and/or business owner will need hand tools and a flashlight or drop light to gain access and to see into the refrigeration appliances.

The chapter will not cover diagnostics, any electrical repairs, or replacement procedures of any sealed system components. This is a specialized area, and only an EPA-certified technician can repair the sealed system on residential refrigeration appliances. The certified technician has the proper training, tools, and equipment required to make the necessary repairs to the sealed system. For the new student, the information in this chapter will be of great importance in your studies to become a certified refrigeration technician. An individual who is not certified to service the sealed system could raise the risk of personal injury, as well as property damage. The uncertified individual could void the manufacturer's warranty on the product if he or she attempts entry into the sealed system.

Residential Refrigerators and Freezers

One of the most important applications of refrigeration (which was invented in the early 1900s) was for the preservation of food. When different types of foods[1] are kept at room temperature, some of them will spoil rapidly. When foods are kept cold, they will last longer. Refrigerators prevent food spoilage by keeping the food cold.

The refrigerator consists of three parts:

1. The cabinet
2. The sealed system, which consists of the evaporator coil, the condenser coil, the compressor, and the connecting tubing
3. The electrical circuitry, including fan motors and other electrical components

This chapter covers the preventive maintenance procedures of domestic refrigerators and freezers. The actual repair or replacement of any sealed-system component is not included in this chapter. It is recommended that you acquire refrigerant certification (or call an authorized service company) to repair or replace any sealed-system component. The refrigerant in the sealed system must be recovered properly.

Safety First

Any person who cannot use basic tools or follow written instructions should *not* attempt to install, maintain, or repair any residential refrigeration appliances. Any improper installation, preventive maintenance, or repairs could create a risk of personal injury or property damage. If you do not fully understand the preventive maintenance procedures in this chapter, or if you doubt your ability to complete the task on your residential refrigeration appliance, please call your appliance service professional.

This chapter covers the preventive maintenance procedures and does not cover how to diagnose the sealed system. The actual repair or replacement of any sealed-system component is not included in this chapter. It is recommended that you acquire refrigerant certification (or call an authorized service company) to repair or replace any sealed-system component, as the refrigerant in the sealed system must be recovered properly.

Before continuing, take a moment to refresh your memory on the safety procedures in Chapter 1.

Residential Refrigeration Appliances in General

Much of the preventive maintenance information in this chapter covers residential refrigeration appliances in general, rather than specific models, in order to present a broad overview of preventive maintenance techniques. The illustrations that are used in this chapter are for demonstration purposes only, to clarify the description of how to perform preventive maintenance on these appliances. They in no way reflect on a particular brand's reliability.

The Refrigeration Cycle

The sealed system (Figure 11-1) in a refrigerator or freezer consists of a compressor, a condenser coil, an evaporator coil, a capillary tube, and a heat exchanger and its connecting tubing. This is the heart of the refrigerator or freezer that keeps the food cold inside the cabinet.

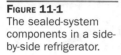

FIGURE 11-1
The sealed-system components in a side-by-side refrigerator.

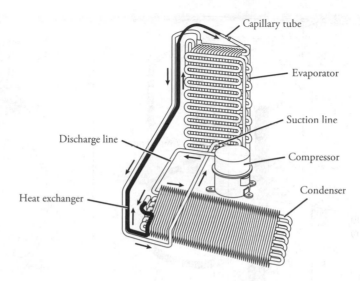

Capillary tube

Evaporator

Suction line

Discharge line

Compressor

Condenser

Heat exchanger

Starting at the compressor, refrigerant gas is pumped out of the compressor, through the discharge tubing, and into the condenser coil. When the gas is in the condenser coil, the temperature and pressure of the refrigerant gas is increased. From the surface of the condenser coil, the heat spreads out into the room via air moving over the condenser coil. The condenser coil cools the hot refrigerant gas. As the refrigerant gas gives up the heat it obtained from inside the refrigerator cabinet, the refrigerant gas changes into a liquid. This liquid then leaves the condenser coil and enters the capillary tube.

This capillary tube is carefully made with regard to its length and inside diameter to meter out the exact amount of liquid refrigerant through the sealed system (this is designed by the manufacturer for a particular size and model). As the liquid refrigerant leaves the capillary tube and enters the larger tubing of the evaporator coil, the sudden increase of tubing size causes a low-pressure area. It is here that the liquid refrigerant changes from a liquid to a mixture of liquid and gas.

As this mixture passes through the evaporator coil, the refrigerant absorbs heat from the warmer items (food) within the refrigerator cabinet, slowly changing any liquid back to gas. As the refrigerant gas leaves the evaporator coil, it returns to the compressor through the suction line.

This entire procedure is called a *cycle*. Depending on where the cold control (thermostat) is set, the thermostat can show how cold it is inside the cabinet and then control the actuation of the cooling cycle. It will determine whether to turn the system on or off to maintain the temperature within the cabinet.

Inside the cabinet, the cold air is circulated by convection and/or by means of an electrical fan. In Figure 11-2, the arrows show the airflow patterns in this type of side-by-side refrigerator. Figure 11-3 shows the air patterns in a two-door refrigerator with a top freezer.

Refrigerator and Freezer Automatic Defrost Systems

Over the years manufacturers have developed many methods to defrost the evaporator coil. Maintaining proper moisture control within the refrigerator or freezer is crucial for proper efficient operation. Every time the consumer opens the door, they allow moisture and heat

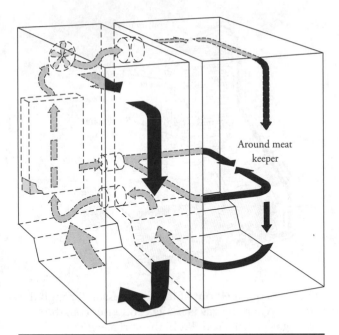

Figure 11-2   The airflow pattern in a side-by-side refrigerator.

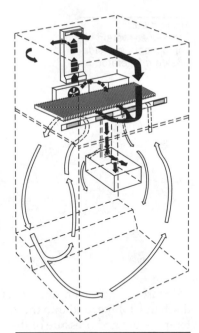

Figure 11-3   The airflow pattern in a two-door refrigerator with top freezer.

to enter the refrigerator or freezer compartment. Through either natural convection or forced-air movement, the moisture will eventually condense on the coldest spot in the refrigerator or freezer: the evaporator coil. The evaporator coil is well below the freezing temperatures, and the condensed moisture will stick to it and begin to freeze. As frost accumulates on the evaporator coil, the airflow through the evaporator coil will begin to decrease and the cooling effect will also decrease.

The manual method for defrosting an evaporator coil will be time-consuming, and most consumers end up destroying the evaporator coil by punching a hole in it with an ice pick or a knife. Manufacturers have developed the automatic defrost system. Most modern-day refrigerators and frost free freezers have a timer to activate the defrost cycle. An electric heater element and a defrost termination thermostat have also been attached to the evaporator coil to defrost the frost buildup. When the timer calls for the defrost cycle, the sealed system is de-energized and the electric heater is energized. The defrost termination thermostat monitors the temperature of the evaporator coil; around 40°F to 60°F the thermostat will turn off the electric heater. The defrost timer will stay in the defrost mode for up to 30 minutes, regardless if the frost is melted or not. The water that results from the defrost process will be directed to the evaporator drain pan and then redirected to the outside of the refrigerator or freezer to the condensate pan at the base of the refrigerator or freezer to be evaporated.

Some refrigerator/freezer models use a defrost cycle on a timed schedule. That means approximately every 6 to 8 hours, the defrost cycle is turned on for approximately 28 to 30 minutes. This type of defrosting system runs without regard for the actual cooling demand on the sealed system. A more efficient type of defrost system developed by manufacturers is the cumulative run-time defrost system. This type of system is based on the compressor run time. The defrost timer only advances when the compressor run time has equaled a predetermined amount of run time; then the system will enter into a defrost time. One disadvantage of this type of defrost system is it does not account for the number of times the door is opened and the increased humidity enters the refrigerator/freezer compartment.

Manufacturers over time have developed a new type of automatic defrost system: the adaptive automatic defrost system. With the decreased cost of electronic components a better way has been developed to control the defrost frequency. This type of system measures the time it takes from the start of the defrost cycle until the defrost termination thermostat opens. This type of system is known as adaptive defrost.

Depending on the manufacturer, the first defrost time will occur between 6 and 8 hours of cumulative compressor run time. The adaptive defrost control (ADC) will continually adjust the defrost intervals based on the following:

- Number of door openings
- Compressor run time
- The last defrost cycle

During the defrost cycle, the ADC monitors how long the defrost termination thermostat keeps the electric heater energized. If the defrost termination thermostat opens in under 12 minutes, this is equal to a light frost build-up. The ADC will increase the amount of compressor run time between defrost times by 2 hours. If the defrost termination thermostat opens in over 12 minutes, this is equal to a heavier frost build-up. The ADC will decrease the amount of compressor run time between defrost times by 2 hours.

Over the course of several defrost cycles, the time between the start of the defrost cycle and the opening of the defrost termination thermostat will get closer to the ideal defrost time. The ADC has adjusted the amount of run time between defrosts to occur often enough to maintain a clean evaporator coil, but not so often as to use excessive energy. As humidity conditions change and the frost load increases or decreases, the ADC will adjust the cumulative run time to match the change in the frost load.

As mentioned earlier, not all refrigerator/freezer manufacturers use this process. They use a slightly different algorithm programmed into the ADC computer chip, but the principle of operation for most of these ADC controls is very similar. The service technician will need the wiring diagram and/or the service manual for the model they are servicing to properly diagnose and be able to turn on the defrost mode on the ADC board.

Another type of defrost system is the hot gas defrost system. This type of system is seldom used in residential refrigerators/freezers due to the additional cost of components. The hot gas defrost systems are used in commercial ice machines to aid in the harvesting process of ice. In the residential refrigerator or freezer, a bypass valve is added to the sealed system to the condenser at a point before the refrigerant begins to condense into a liquid state. When defrost is called for, the valve will open, and the hot gas enters the evaporator after the capillary tube inlet. The hot gas will thaw the frost from the evaporator coil.

A cycle defrost system is used on inexpensive refrigerators. The freezer evaporator is manual defrost, but the refrigerator evaporator is defrosted every time the compressor cycles off. In this type of system, a special thermostat and two low-wattage electric heaters are used to defrost the evaporator.

Storage Requirements for Perishable Products

Table 11-1 lists the recommended storage temperatures, relative humidity, and the approximate storage life for perishable products. These values are used in designing commercial refrigeration systems, which house large quantities of perishable products. Large warehouses are usually equipped to store foods at those temperatures best adapted to prolonging the safe storage period for each type of food. In the domestic refrigerator[1], most foods are kept at 34°F to 40°F (1°C to 4°C) with an optimum temperature of 37°F, and the humidity is kept around 50%. The freezer temperature is between 0°F and –10°F (–17°C to –23°C).

Product	Storage Temp.°F	Relative Humidity %	Approximate Storage Life
Apples	30–40	90	3–8 months
Apricots	31–32	90	1–2 weeks
Artichokes	31–32	95	2 weeks
Asparagus	32–36	95	2–3 weeks
Avocados	45–55	85–90	2–4 weeks
Bananas	55–65	85–95	—
Beans (green or snap)	40–45	90–95	7–10 days
Beans, lima	32–40	90	1 week
Blackberries	31–32	95	3 days
Blueberries	31–32	90–95	2 weeks
Broccoli	32	95	10–14 days
Cabbage	32	95–100	3–4 months
Carrots	32	98–100	5–9 months
Cauliflower	32	95	2–4 weeks
Celery	32	95	1–2 months
Cherries, sour	31–32	90–95	3–7 days
Cherries, sweet	30–31	90–95	2–3 weeks
Collards	32	95	10–14 days
Corn, sweet (fresh)	32	95	4–8 days
Cranberries	36–40	90–95	2–4 months
Cucumbers	50–55	90–95	10–14 days
Dairy Products			
Cheddar cheese	40	65–70	6 months
Processed cheese	40	65–70	12 months

TABLE 11-1 Storage Requirements for Perishable Products for Commercial Refrigeration and Freezers (*continued*)

Product	Storage Temp.°F	Relative Humidity %	Approximate Storage Life
Butter	40	75–85	1 month
Cream	35–40	—	2–3 weeks
Ice cream	–20 to –15	—	3–12 months
Milk, fluid whole			
Pasteurized, grade A	32–34	—	2–4 months
Condensed, sweetened	40	—	15 months
Evaporated	40	—	24 months
Dates (dried)	0 or 32	75 or less	6–12 months
Dried fruits	32	50–60	9–12 months
Eggplant	45–50	90–95	7–10 days
Eggs, shell	29–31	80–85	5–6 months
Figs, dried	32–40	50–60	9–12 months
Figs, fresh	31–32	85–90	7–10 days
Fish, fresh	30–35	90–95	5–15 days
Haddock, cod	30–35	90–95	15 days
Salmon	30–35	90–95	15 days
Smoked	40–50	50–60	6–8 months
Shellfish, fresh	30–33	86–95	3–7 days
Tuna	30–35	90–95	15 days
Grapefruit	50–60	85–90	4–6 weeks
Grapes, American type	31–32	85–90	2–8 weeks
Grapes, European type	30–31	90–95	3–6 months
Greens, leafy	32	95	10–14 days
Guavas	45–50	90	2–3 weeks
Honey	38–50	50–60	1 year, plus
Horseradish	30–32	95–100	10–12 months
Lemons	32 or 50–58	85–90	1–6 months
Lettuce, head	32–34	95–100	2–3 weeks
Limes	48–50	85–90	6–8 weeks
Maple sugar	75–80	60–65	1 year, plus
Mangoes	55	85–90	2–3 weeks
Meat			
Bacon, cured (farm style)	60–65	85	4–6 months
Game, fresh	32	80–85	1–6 weeks
Beef, fresh	32–34	88–92	1–6 weeks

TABLE 11-1   Storage Requirements for Perishable Products for Commercial Refrigeration and Freezers (*continued*)

Product	Storage Temp.°F	Relative Humidity %	Approximate Storage Life
Hams and shoulders, fresh	32–34	85–90	7–12 days
Cured	60–65	50–60	0–3 years
Lamb, fresh	32–34	85–90	5–12 days
Livers, frozen	–10–0	90–95	3–4 months
Pork, fresh	32–34	85–90	3–7 days
Smoked sausage	40–45	85–90	6 months
Fresh	32	85–90	1–2 weeks
Veal, fresh	32–34	90–95	5–10 days
Melons, Cantaloupe	36–40	90–95	5–15 days
Honeydew and Honey Ball	45–50	90–95	3–4 weeks
Watermelons	40–50	80–90	2–3 weeks
Mushrooms	32	90	3–4 days
Milk	34–40	—	7 days
Nectarines	31–32	90	2–4 weeks
Nuts (dried)	32–50	65–75	8–12 months
Okra	45–50	90–95	7–10 days
Olives, fresh	45–50	85–90	4–6 weeks
Onions (dry) and onion sets	32	65–70	1–8 months
Oranges	32–48	85–90	3–12 weeks
Orange juice, chilled	30–35	—	3–6 weeks
Papayas	45	85–90	1–3 weeks
Parsley	32	95	1–2 months
Parsnips	32	98–100	4–6 months
Peaches	31–32	90	2–4 weeks
Pears	29–31	90–95	2–7 months
Peas, green	32	95	1–3 weeks
Peppers, sweet	45–50	90–95	2–3 weeks
Pineapples, ripe	45	85–90	2–4 weeks
Plums, including fresh prunes	31–32	90–95	2–4 weeks
Popcorn, unpopped	32–40	85	4–6 months
Potatoes, early crop	50–55	90	0–2 months
Potatoes, late crop	38–50	90	5–8 months
Poultry			
Fresh chicken	32	85–90	1 week
Fresh goose	32	85–90	1 week

TABLE 11-1   Storage Requirements for Perishable Products for Commercial Refrigeration and Freezers (*continued*)

Product	Storage Temp.°F	Relative Humidity %	Approximate Storage Life
Fresh turkey	32	85–90	1 week
Pumpkins	50–55	70–75	2–3 months
Radishes—spring, prepacked	32	95	3–4 weeks
Raisins (dried)	40	60–70	9–12 months
Rabbits, fresh	32–34	90–95	1–5 days
Raspberries, black	31–32	90–95	2–3 days
Raspberries, red	31–32	90–95	2–3 days
Rhubarb	32	95	2–4 weeks
Spinach	32	95	10–14 days
Squash, summer	32–50	85–95	5–14 days
Squash, winter	50–55	70–75	4–6 months
Strawberries, fresh	31–32	90–95	5–7 days
Sugar, maple	75–80	60–65	1 year, plus
Sweet potatoes	55–60	85–90	4–7 months
Syrup, maple	31	60–70	1 year, plus
Tangerines	32–38	85–90	2–4 weeks
Tomatoes, mature green	55–70	85–90	1–3 weeks
Tomatoes, firm ripe	45–50	85–90	4–7 days
Turnips, roots	32	95	4–5 months
Vegetables (mixed)	32–40	90–95	1–4 weeks
Yams	60	85–90	3–6 months

TABLE 11-1   Storage Requirements for Perishable Products for Commercial Refrigeration and Freezers

It can be difficult to maintain these temperatures and humidity for each individual product. Therefore, refrigerator manufacturers have designed separate compartments within the refrigerated cabinet to maintain a variable temperature and humidity selected by the consumer. The storage life of various products will vary in a domestic refrigerator/freezer (Table 11-2). This period will be influenced by many factors, such as the storage temperature, the type of container, the condition of the food, and the kind of food. For food storage tips consult the use and care guide that comes with the appliance.

To test the temperature in a refrigerator/freezer, place a thermometer in a glass of water and place it in the center of the refrigerator compartment. After 24 hours, check the thermometer. To test the temperature in the freezer compartment, place a thermometer between two frozen packages. After 24 hours, check the thermometer. If the temperature controls need to be readjusted, retake temperatures as listed above.

Domestic Refrigerator/Freezer Food Storage Tips			
Foods	**Refrigerator**	**Freezer**	**Storage Tips**
Dairy Products			
Butter	1 month	6–9 months	Wrap product tightly or cover.
Cream cheese, cheese spread, and cheese food	1–2 weeks	Not recommended	Wrap product tightly.
Cottage cheese	3–5 days	Not recommended	Keep product stored in original carton. Check carton date.
Hard cheese (Swiss, cheddar, and parmesan)	1–2 months	4–6 months May become crumbly	Wrap tightly. Cut off mold.
Milk and cream	1 week	Not recommended	Check the date on the carton. Close tightly. Do not store unused portions in the original container. Do not freeze the cream unless whipped.
Sour cream	10 days	Not recommended	Keep product stored in original carton. Check carton date.
Eggs			
Eggs in the shell	3 weeks	Not recommended	Refrigerate with the small ends facing down.
Leftover yolks or whites	2–4 days	9–12 months	For each cup of yolks to be frozen, add 1 teaspoon of sugar for use in sweet dishes, or 1 teaspoon of salt for nonsweet dishes.
Fruits			
Apples	1 month	8 months (cooked)	Store unripe or hard apples at 60°–70°F (16°–21°C).
Bananas	2–4 days	6 months (whole peeled)	Ripen at room temperature before refrigerating. Bananas will darken when refrigerated.
Berries, cherries, apricots	2–3 days	6 months	Ripen at room temperature before refrigerating.
Citrus fruits	1–2 weeks	Not recommended	Store uncovered at 60°–70°F (16°–21°C).
Grapes	3–5 days	1 month (whole)	Ripen at room temperature before refrigerating.
Pears, plums, avocados	3–4 days	Not recommended	Ripen at room temperature before refrigerating. Avocados will darken when refrigerated.

TABLE 11-2   Storage Requirements for Perishable Products for a Domestic Refrigerator/Freezer (*continued*)

Domestic Refrigerator/Freezer Food Storage Tips			
Foods	**Refrigerator**	**Freezer**	**Storage Tips**
Pineapples, cut	2–3 days	6–12 months	Will not ripen after purchase. Use quickly.
Vegetables			
Asparagus	1–2 days	8–10 months	Do not wash before refrigerating. Store in crisper drawer.
Brussels sprouts, broccoli, cauliflower, green peas, lima beans, onions, peppers	3–5 days	8–10 months	Wrap odorous foods. Leave the peas in the pods.
Cabbage, celery	1–2 weeks	Not recommended	Wrap odorous foods and refrigerate in crisper drawer.
Carrots, parsnips, beets, turnips	7–10 days	8–10 months	Remove tops. Wrap odorous foods and refrigerate in crisper drawer.
Lettuce	7–10 days	Not recommended	
Poultry and Fish			
Chicken and turkey, pieces	1–2 days	9 months	Keep in original packaging for refrigeration. Place in the meat and cheese drawer. When freezing longer than 2 weeks, overwrap with freezer wrap paper.
Chicken and turkey, whole	1–2 days	12 months	
Fish	1–2 days	2–6 months	
Meats			
Bacon	7 days	1 month	
Beef or lamb, ground	1–2 days	3–4 months	Fresh meats can be kept in original packaging for refrigeration.
Beef or lamb, roast and steak	3–5 days	6–9 months	Place in the meat and cheese drawer. When freezing longer than 2 weeks, overwrap with freezer wrap paper.
Frankfurters	7 days	1 month	Processed meats should be tightly wrapped and stored in the meat and cheese drawer.
Ham, fully cooked, half	5 days	1–2 months	
Ham, fully cooked, slices	3 days	1–2 months	
Ham, fully cooked, whole	7 days	1–2 months	

TABLE 11-2   Storage Requirements for Perishable Products for a Domestic Refrigerator/Freezer (*continued*)

Domestic Refrigerator/Freezer Food Storage Tips			
Foods	**Refrigerator**	**Freezer**	**Storage Tips**
Pork, chops	3–5 days	4 months	
Pork, roast	3–5 days	4–6 months	
Sausage, ground	1–2 days	1–2 months	
Veal	3–5 days	4–6 months	
Luncheon meats	3–5 days	1–2 months	Unopened, vacuum-packed luncheon meat may be kept up to 2 weeks in the meat and cheese drawer.

TABLE 11-2   Storage Requirements for Perishable Products for a Domestic Refrigerator/Freezer

Storing Foods in a High Humidity Environment within a Domestic Refrigerator

When storing unwrapped foods in a high humidity environment, such as the crisper drawers located in the bottom of a refrigerator, that have individual adjustable humidity controls that allows the consumer to control the amount of cold air entering the drawers, keeps the food fresh by retaining the natural moisture content of foods such as:

- Artichokes
- Asparagus
- Beets
- Blueberries
- Carrots
- Celery
- Cherries
- Corn
- Cucumber
- Currants
- Greens, leafy
- Lettuce
- Parsley
- Peas, green
- Radishes
- Rhubarb
- Spinach
- Tomatoes

As in any refrigerated storage area, it is recommended that foods with strong odors be stored wrapped, such as:

- Broccoli
- Brussels sprouts
- Cabbage
- Cauliflower
- Green onions
- Parsnips
- Turnips

Storing Foods in a Low Humidity Environment within a Domestic Refrigerator

Storing foods in a low humidity environment, such as the crisper drawers located in the refrigerator, which have individual adjustable humidity controls that allows the consumer to control the amount of cold air entering the drawers, can be used for foods such as:

- Apples
- Apricots
- Grapes
- Mushrooms
- Nectarines
- Oranges
- Papayas
- Peaches
- Pears
- Pomegranates
- Mangoes
- Raspberries
- Squash
- Strawberries
- Tangerines

Storage Requirements and Serving Requirements for Wines

Table 11-3 indicates the storage temperature of wines.

Type of Wine	Serving Temperature °F
Champagne and Other Sparkling Wines	
Better non-vintage	42–47
Best vintage	45–50
Inexpensive versions	42–44
Sweet and Semi-Sweet White Wines	
Dessert wines	45–50
German semi-sweet wines	45–50
German older sweet wines	50–55
Sauternes (older)	50–55
Vouvray	43–48
White Wines (Dry and Off-Dry)	
Bordeaux	45–50
Burgundy medium quality	45–50
Burgundy premium quality	50–55
Chardonnay medium quality	45–50
Chardonnay premium quality	50–55
Chenin Blanc U.S.	43–48
German high quality	48–53
Gewurztraminer	45–50
Italian	45–50
Melon de Bourgogne	43–48
Muscadet	43–48
Pinot Blanc	45–50
Pinot Gris	45–50
Pouilly-Fume	43–48
Rhone northern	48–53
Rhone southern	45–50
Sancerre	43–48
Sauvignon Blanc	43–48
Savenniers	43–48
Rose and Blush Wines	
Almost all rose and blush wines	40–45
Dry rose (high quality)	43–48

TABLE 11-3 Storage Temperatures of Wines (*continued*)

Type of Wine	Serving Temperature °F
Red Wines	
Aglianico wines from Southern Italy	62–67
Barbaresco	60–65
Barbera U.S. and Italian	57–62
Bardolino	48–53
Barolo	62–67
Beaujolais	50–55
Bordeaux most vintages	58–63
Bordeaux best vintages	62–67
Brunello di Montalcino	62–67
Burgundy better vintages	58–63
Burgundy most Cote de Beaune	55–60
Burgundy most Cote de Nuits	58–63
Cabernet Sauvignon U.S.	62–67
Chianti Classico	57–62
Chianti Classico Reserva	60–65
Cotes du Rhone	53–58
Dolcetto U.S. and Italian	53–58
German Red Wines	52–57
Lambrusco	48–53
Merlot U.S.	59–64
Nouveau style French and U.S.	48–53
Pinot Noir U.S.	58–63
Rhone	62–67
Rioja	57–62
Rioja Reserva	60–65
Shiraz, Australian	59–64
Syrah U.S.	59–64
Valpolicella Amarone style	55–60
Valpolicella except Amarone style	52–57
Zinfandel light	57–62
Zinfandel heavy	62–67

TABLE 11-3 Storage Temperatures of Wines (*continued*)

Type of Wine	Serving Temperature °F
Fortified Wines	
Madeira	55–60
Muscat de Beaumes de Venise	42–47
Port, Ruby	57–62
Port, Tawny older	55–60
Port, Tawny younger	50–55
Port, Vintage	62–67
Port, white	45–50
Sherry, amontillados	60–65
Sherry, most finos	45–50
Sherry, olorosos	60–65

TABLE 11-3   Storage Temperatures of Wines

Location and Installation of Refrigerator/Freezer

Thoroughly read the installation instructions that are provided with every new refrigerator or freezer. These instructions will provide you with the information you need to properly install the refrigerator or freezer. The following are some general principles that should be followed when performing the installation:

- The refrigerator or freezer must be installed on a solid floor capable of supporting the product up to 1000 lb.
- For proper air circulation around the refrigerator/freezer, some models require a 1-in clearance at the rear and top of the cabinet and adequate clearance near the front grille at the bottom of the refrigerator.
- Do not leave the refrigerator/freezer on its side longer than necessary to remove the shipping base.
- When removing or reversing the doors on a refrigerator or freezer, always reinstall them according to the installation instructions, and remember to realign the doors properly.
- Level the refrigerator or freezer cabinet so that the doors close properly.

Refrigerator/Freezer Maintenance

Before starting the maintenance on the refrigerator/freezer, turn off the electricity to the appliance. Start from top to bottom and remove the food as needed to begin the maintenance. The inside of the cabinet should be cleaned at least once a month to help prevent odors from building up. Of course, any spills that might happen should be wiped up immediately. Wash all removable parts by hand with warm water and a mild detergent; then rinse and dry the parts. The inside walls of the cabinet, the door liners, and the gaskets should also be washed using warm water and a mild detergent, rinsed, and dried. Apply a thin layer of petroleum jelly on the door gaskets at the hinge side. This will help keep the gaskets from sticking and bending out of shape. If the door gaskets are ripped or torn, replace as necessary, to prevent the cold air spilling out of the appliance.

On refrigerator/freezers, with bad odor problems, you can use baking soda, about a tablespoon (15 mL) mixed with a quart (1 L) of warm water. Never use cleaning waxes, concentrated detergents, bleaches, ammonia, or cleansers containing petroleum products on plastic parts. Never use cleaning products with a lemon scent; the lemon will be absorbed into the liner permanently and it may also affect the food. On the outside of the cabinet, use a sponge with warm water and a mild detergent to clean dust and dirt. Then rinse off and dry thoroughly.

At least two times a year, the outside cabinet should be waxed with an appliance wax or with a good auto paste wax. Waxing painted metal surfaces provides rust protection. The defrost pan, which is located behind the toe plate or behind the cabinet, should be cleaned out once a month. The condenser coil should also be cleaned of dust and lint at least once a month in a dusty environment and 2 to 3 months in a normal dust-free environment. In homes or business with pets, the condenser coil might have to be cleaned monthly to ensure maximum efficiency. The floor should be free of dirt and debris when the cabinet is rolled out away from the wall. After the cabinet is rolled back into place, you must check to be sure that the cabinet is level. Read the manufacturer's use and care guide that came with the appliance for any additional preventive maintenance that must be performed on your model appliance.

SAFETY NOTE *Some refrigerators/freezers contain R600a (isobutene) refrigerant. R600a is a highly combustible refrigerant. When handling the refrigerator/freezer, installing, or operating the appliance extreme care must be taken to avoid damage to the refrigerant tubing. If your refrigerator/freezer has R600a refrigerant, read the literature that came with the appliance from the manufacturer for additional safety and handling warnings.*

Food Odors and Molds

Odors in the refrigerator compartment or the freezer compartment cannot occur by themselves. The only way that odors can occur is by storing foods in an unsealed container or unwrapped. Another way that odors occur is from food spillage or from rotten or spoiled food. In new refrigerators there may be a plastic odor, but this is normal and it will dissipate in time. Here are some more tips to help in odor removal:

- Place a box of baking soda in the refrigerator's fresh food and freezer compartments. Replace according to the instructions on the box.
- Place some activated charcoal in a shallow metal pan inside the fresh food or the freezer compartment of the refrigerator. When the charcoal loses its effectiveness, place the metal pan in the oven and heat it on a low temperature for a couple of

hours to rejuvenate it. Do not use charcoal briquettes used for grilling; it is not the same activated charcoal.

- Place some vanilla extract in a small dish and place in the refrigerator's fresh food compartment for 3 weeks. Do not place in the freezer compartment; the vanilla extract will freeze and be ineffective.

On occasion the ice cubes will have a bad taste and will smell like food. Sometimes, the food odors come from the refrigerator compartment. To be able to tell which storage compartment is producing the food odors, try the following:

- Fill ice trays with tap water and freeze them.
- Remove the ice cubes from the tray and place in a bowl.
- Place a bowl in the freezer compartment for a few days.
- Taste or smell the ice cubes.

If the taste or odor is present in the ice cubes, then the odors are coming from the refrigerator or freezer compartment. The odors are present in the air and as they circulate between the two compartments, the ice cubes absorb the odors. If the bowl of ice cubes has no odors or bad taste, then the bad taste and odor are coming from the water supply that feeds the automatic ice maker and/or water dispenser. Once every 2 or 3 weeks, replace the old ice cubes with new ones. Some model refrigerators have a water filter installed for the ice cube maker and water dispenser, this filter has to be replaced every 6 months, for good water quality.

Food molds often grow on baked goods, produce, and leftovers and dairy products. Mold is caused by microbes that attach to the food surface and causes the food to go bad. Underneath the food's surface the mold cells attack the remainder of the food, causing it to rot out.

As food is stored in the refrigerator it loses taste, texture, and nutritional value. Improper handling or storing of foods can cause food-related illness or even disease. To clean the mold from the refrigerator, throw away spoiled food, and clean the walls and shelves in the refrigerator and freezer compartments. Follow the use and care instructions that came with the refrigerator for proper cleaning instructions.

Freezers

Freezers are conveniences for people who have very large families or for people who do not frequent the supermarket. They are especially useful in homes with smaller refrigerators or refrigerators having only an ice cube tray compartment. Home freezers come in chest and upright models. Two designs of upright models are available on the market today: manual defrost and automatic defrost. Home freezers are available with wire shelves and baskets, and with storage shelves on the doors in upright models.

The sealed system in the freezer operates the same as the refrigerator/freezer models. The only difference is temperature. The domestic freezer operates at a colder temperature. The reason for colder temperatures is to maintain food preservation for a longer period of time.

Upright Freezers

Upright freezers are similar to refrigerator/freezers in design and operation. They share some of the same features:

- Fan motors
- Compressor
- Automatic defrost system
- Door gasket
- Thermostat
- Interior lighting

On manual-defrost models,[2] the evaporator coils are the shelves inside the cabinet. Figure 11-4 illustrates the refrigerant flow on this type of manually defrosted upright freezer. The condenser coils are embedded between the cabinet liners and are secured to the inside wall of the outer cabinet. This provides for even heat removal and it eliminates the need for a condenser fan motor.

Automatic-defrost models[3] use a fan motor to circulate the air inside the cabinet through air ducts. The evaporator coil is mounted on the inside back wall of the inner liner. Figure 11-5 illustrates the airflow pattern in an upright automatic-defrost freezer. Figure 11-6 illustrates the refrigerant flow in this type of upright freezer.

Chest Freezers

The chest freezer (Figure 11-7) has the evaporator coils and the condenser coils embedded between the inner liner and the outer cabinet. These coils are inaccessible for replacement or

FIGURE 11-4
Refrigerant flow of an upright freezer with manual defrost.

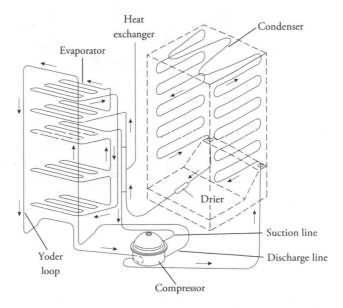

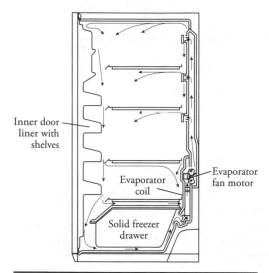

FIGURE 11-5   The airflow pattern of an upright automatic-defrost freezer.

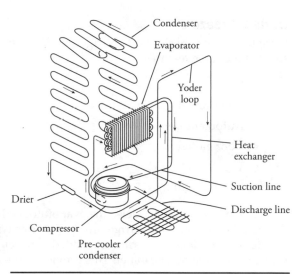

FIGURE 11-6   The refrigerant flow of an upright automatic-defrost freezer.

repair if a refrigerant leak occurs. The differences between the upright freezer and the chest freezer are

- Door hinges
- Gasket

FIGURE 11-7
The component location of a chest freezer.

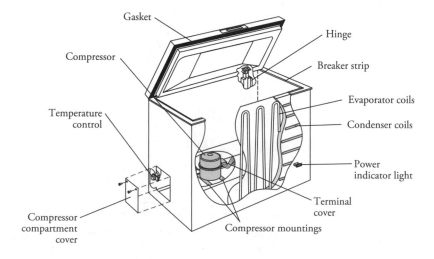

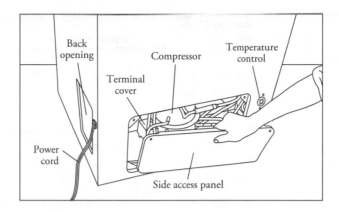

FIGURE 11-8
Removing the side access panel to gain access to the components.

Back opening

Compressor

Temperature control

Terminal cover

Power cord

Side access panel

- Location and access of temperature controls
- Location and access of the compressor, relay, and overload protector

Most models have a power indicator light. This light stays on as long as the freezer is plugged into the wall receptacle. The light alerts the consumer when the power is off to the freezer, but it does not tell you what the temperature is inside the cabinet.

Chest freezers must be defrosted once or twice a year to remove the ice buildup from the inside. To gain access to the components, remove the side access panel (Figure 11-8).

Today some manufacturers have designed chest freezers with an automatic defrost feature similar to upright freezers.

Freezer Storage and Temperatures

Storing and preserving food in a freezer is a great way to maintain the quality, freshness, and nutritional value of food products. The recommended freezer temperature should be 0°F or colder. At this temperature food should last indefinitely if packaged properly. Remember these three rules when packaging foods for freezer storage: wrap tightly, double-wrap the product again, and wrap individual portions separately. To prevent freezer burn, you must wrap food tightly to remove as much air as possible and also double-wrap the food product. When you wrap food in individual portions, you do not have to thaw out a large amount of food. You just take out what you need for that meal.

Maintaining a food's quality depends on several factors: the quality of the raw product, the procedures used during processing, the way the food is stored, and the length of storage. The recommended storage time takes these factors into consideration.

Table 11-4 depicts a storage chart for various types of food products. Also, I recommend that dates be placed on food products to protect the consumer from eating spoiled or outdated food.

Food	Storage Time at 0°F	Comments
Meat, Fish, Poultry		
Bacon	Freezing not recommended	Saltiness encourages rancidity.
Corned beef	Use within 1 month	
Frankfurters	1–2 months	Emulsion may be broken and the product will "weep."
Ground beef, lamb, or veal	2–4 months	
Ground pork	1–2 months	
Ham and picnic cured ham	1–2 months	Saltiness encourages rancidity.
Luncheon meats	Use within 1 month	Emulsion may be broken and the product will "weep."
Beef roast	4–12 months	Freeze product in original packaging for 2 weeks. If needed to store longer, wrap in freezer wrap. For patties, separate with wax paper.
Lamb or veal roast	6–9 months	
Pork roast	3–6 months	
Sausage, dry, smoked	Use within 1 month	Freezing alters the flavor.
Beef steaks	6–9 months	Freeze product in original packaging for 2 weeks. If needed to store longer, wrap in freezer wrap. For patties, separate with wax paper.
Beef chops	4–6 months	
Lamb or veal steaks and chops	3–4 months	
Pork steaks and chops	2–3 months	
Venison or game birds	6–12 months	
Fish		
Bluefish, mackerel, perch, and salmon	2–3 months	Freeze product in original packaging for 2 weeks. If needed to store longer, wrap in freezer wrap.
Breaded fish	3 months	
Clams	2–3 months	
Cooked fish or seafood	3–6 months	
King crab	5 months	
Lean fish, cod, flounder, haddock, and sole fillets and steaks	6 months	
Lobster tails	2–3 months	
Oysters	2–4 months	
Scallops	3–6 months	
Shrimp, uncooked	10 months	

TABLE 11-4   Appropriate Freezing Periods for Various Types of Foods (*continued*)

Food	Storage Time at 0°F	Comments
Poultry		
Chicken, cut up or whole	9–12 months	Freeze product in original packaging for 2 weeks. If needed to store longer, wrap in freezer wrap. For patties, separate with wax paper.
Chicken livers	3–4 months	
Cooked poultry	3–5 months	
Duck, turkey	6–9 months	
Fruit		
Berries, cherries, peaches, pears, pineapples, etc.	12 months	
Citrus fruit and juice frozen at home	6 months	
Fruit juice concentrates	8–12 months	
Vegetables		
Home frozen	10 months	Cabbage, celery, salad greens, and tomatoes for slicing do not freeze successfully; tomatoes for soups, stews, or sauces can be frozen successfully.
Purchased frozen	8 months	
Baked Goods		
Yeast bread and rolls, baked	2–4 months	Freezing does not refresh the baked goods. It can only maintain whatever the quality of the food was before freezing.
Rolls, partially baked	2–3 months	
Bread, unbaked	1 month	
Quick bread, baked	2–3 months	
Cake, baked, unfrosted	2–3 months	
Angel food cake	2–6 months	
Chiffon sponge cake	2 months	Freezing does not refresh the baked goods. It can only maintain whatever the quality of the food was before freezing.
Cheesecake	2–3 months	
Chocolate cake	4 months	
Fruit cake	6–12 months	
Yellow or pound cake	6 months	
Cake, baked, frosted	3–4 months	
Cookies, baked	8–12 months	
Pie, baked	1–2 months	
Fruit pie, baked	6–8 months	
Cake, unbaked	1 month	
Main Dishes		
Meat, fish, and poultry, pies and casseroles	3–4 months	

TABLE 11-4   Appropriate Freezing Periods for Various Types of Foods (*continued*)

Food	Storage Time at 0°F	Comments
TV dinners, including shrimp, ham, pork, or frankfurter	3–4 months	
TV dinners, including beef, turkey, chicken, or fish	6 months	
Frozen Foods—Home		
Baked muffins	6–12 months	
Unfrosted doughnuts	2–4 months	
Waffles	1 month	
Bread	3 months	
Cake	3 months	
Casseroles—meat, fish, or poultry	2–4 months	
Cookies, baked and dough	2–3 months	
Nuts, salted	6–8 months	
Nuts, unsalted	9–12 months	
Pies, unbaked fruit	8 months	
Dairy Products		
Butter	6–9 months	
Margarine	12 months	
Whipped butter and margarine	*Do not freeze*	Emulsion will break and the product will separate.
Buttermilk, sour cream, and yogurt	*Do not freeze*	
Camembert cheese	3 months	
Cottage, farmer's cheese (dry curd only)	1–3 months	Do not freeze creamy cottage cheese; it will become mushy.
Neufchatel cheese	*Do not freeze*	
Cheddar cheese	6 weeks	
Edam, gouda, Swiss, brick cheeses, etc.	6–8 weeks	
Processed cheese food products (loaves, slices)	4–6 months	
Roquefort, blue cheese	3 months	
Cream—light, heavy, half and half	2–4 months	
Whipped cream	1–2 months	

TABLE 11-4   Appropriate Freezing Periods for Various Types of Foods (*continued*)

Food	Storage Time at 0°F	Comments
Eggs in shell	*Do not freeze*	Yolks will thicken and will taste unsatisfactory in cooked products.
Whole eggs out of the shell or egg yolks	12 months	Beat thoroughly with either ½ teaspoon of salt or 2 tablespoons of sugar per cup of yolk or whole egg to control the thickening of the yolk; use in food products that ordinarily use salt or sugar as an ingredient.
Egg whites	12 months	Added salt or sugar not necessary.
Ice cream, ice milk, sherbet, frozen yogurt	2 months	
Milk	1 month	

TABLE 11-4   Appropriate Freezing Periods for Various Types of Foods

Endnotes

1. Dairy products, meats, seafood, fruits, and vegetables will all spoil rapidly if not kept cold or frozen. The colder temperatures of the refrigeration compartment should be between 35°F and 40°F to prevent the spoiling of foods. Most foods will last from 3 to 7 days at that temperature. If the foods were frozen and packaged properly, they could last for several weeks in a domestic refrigerator with a temperature of 0°F to –10°F.

2. Manual defrost freezers must be defrosted once or twice a year, depending upon usage.

3. Automatic defrost freezers will self-defrost the evaporator coil. The defrost components are the defrost timer, the defrost thermostat, and the defrost heater.

Commercial Air Conditioning and Heating

This chapter will only cover the basics needed to perform the preventive maintenance on commercial air conditioning and heating systems. Each preventive maintenance procedure has been rated by a degree of complexity or simplicity. As the owner of the product, you will have to decide whether you can perform the preventive maintenance procedure or if you must call a professional to complete the preventive maintenance procedures on the product. Listed below are symbols that are used in this chapter to define who should perform the preventive maintenance on the air conditioning and heating equipment.

The business owner will need hand tools and a flashlight or drop light to gain access and to see into the HVAC equipment.

Expert   Call a certified HVAC service Company

Hard   Service technician

Moderate   Business Owner

Easy   Home Owner

The chapter will not cover diagnostics, any electrical repairs, or replacement procedures of any sealed-system components. This is a specialized area, and only an EPA-certified technician can repair the sealed system on air conditioning equipment. The certified technician has the proper training, tools, and equipment required to make the necessary repairs to the sealed system. For the new HVAC student, the information in this chapter will be of great importance in your studies to become a certified air conditioning technician. An individual who is not certified to service the sealed system could raise the risk of personal injury, as well as property damage. The uncertified individual could void the manufacturer's warranty on the product if he or she attempts entry into the sealed system.

Commercial Air Conditioning and Heating Equipment

The air conditioning equipment involves many aspects of conditioning the air in a commercial building and/or including introducing fresh air in whatever way necessary to make the working environment for the occupants comfortable. This process may include cooling the air, warming the air, filtering the air, adding moisture to the air, removing moisture from the air, and maintaining a well-balanced air flow and duct-distribution system. Air conditioning includes:

- Cooling the occupied space
- Heating the occupied space
- Humidification (moisture in the air)
- Dehumidification (removing moisture from the air)
- Preventive maintenance on the equipment
- Cleaning the equipment
- Filtration of the recirculating air within the conditioned space
- Humidity control within the occupied space

Principles of Operation

The commercial air conditioning unit, when installed and running properly, will circulate the air through ducts in the room(s) or area, removing the heat and humidity; some models will heat the air by electric, gas, oil, or water in the winter months. At the same time, the air filter, located in the return air grille, will filter out dust particles. Some models have a fresh air intake feature, which allows fresh outside air to enter the return duct when the unit is running. The thermostat will control the comfort level in the room(s) or area, and cycle the HVAC equipment on and off according to the temperature setting.

Safety First

Any person who cannot use basic tools or follow written instructions should *not* attempt to install, maintain, or repair any commercial air conditioning unit. Any improper installation, preventive maintenance, or repairs could create a risk of personal injury or property damage. If you do not fully understand the preventive maintenance procedures in this chapter, or if you doubt your ability to complete the task on the commercial air conditioner, please call your HVAC service professional.

This chapter covers the preventive maintenance procedures and does not cover how to diagnose the sealed system. The actual repair or replacement of any sealed-system component is not included in this chapter. It is recommended that you acquire refrigerant certification (or call an authorized service company) to repair or replace any sealed-system component, as the refrigerant in the sealed system must be recovered properly.

Before continuing, take a moment to refresh your memory on the safety procedures in Chapter 1.

Commercial Air Conditioning and Heating Equipment in General

Much of the preventive maintenance information in this chapter covers commercial air conditioners in general, rather than specific models, in order to present a broad overview of preventive maintenance techniques. The illustrations that are used in this chapter are for demonstration purposes only, to clarify the description of how to perform preventive maintenance on the HVAC equipment. They in no way reflect on a particular brand's reliability.

Preventive Maintenance Checklist

The following checklist will assist the certified HVAC service technician in performing a proper inspection on commercial air conditioning and heating equipment including package units and water source units. By properly maintaining the equipment, the efficiency of the equipment will be operating at or above the 90% range of the rated BTU capacity with a reduction in energy use. Always check the manufacturer's literature that comes with the HVAC equipment if there are additional preventive maintenance checks that are not listed on this checklist.

Air Conditioning Unit

- Check and adjust the thermostat.
- Check to make sure indoor and outdoor unit come on.
- If the HVAC equipment is contained in one unit (package unit), check to make sure all fans and the compressor are running.
- Check sequence of operation.
- Replace air filter or clean reusable type filter.
- Check and inspect the ducts to make sure that there are no air leaks or discrepancies.
- Check the duct grilles and registers for proper air flow.
- Check smoke/fire dampers.
- Check the evaporator coil and advise if dirty or if it needs cleaning.
- Check expansion valve and coil temperatures.
- If the unit you are inspecting has a belt driven blower motor, check the blower belt wear, tension, and adjust as needed.
- Check bearings and lubricate blower motor if needed.
- Check fresh air intake.
- Check the condensate drain system and secondary drain pan then advise of any discrepancies.
- Check condensate drain pan float switches for proper operation and installation.
- Check the condenser coil to determine if it needs cleaning.
- Inspect all wiring and connections to controls and electrical connections.
- Check operation of variable speed frequency drive for operation.

- Check voltage and amperage draw on all motors with a clamp-on multimeter.
- Check compressor contactor, capacitor, and any other controls in the condenser unit or package unit.
- Check the compressor and the amperage draw and compare the reading against the data information plate that is attached to the unit.
- Check all controls.
- Check UVC lamps, and clean and check for proper operation.
- If the unit you are inspecting has a start capacitor and potential relay, check it.
- If the unit you are inspecting has a pressure switch, check the cut-out settings.
- Install refrigerant gauges and check operating pressures, superheat, and subcooling.
- Check refrigerant levels and advise the owner if adjustment or repairs are necessary.
- Check fresh air damper for proper operation. Repair, replace, or adjust as necessary.
- Visually inspect exposed ducts and external piping for insulation and vapor barrier for integrity. Correct any problems found.
- Check integrity of all panels on HVAC unit. Replace missing fasteners as needed to ensure proper integrity and fit of HVAC equipment.
- Check the conditions of the entire air conditioning system and advise the owner of any discrepancies.

Electric Heat

- Check and adjust the thermostat.
- Check and see if the indoor unit comes on.
- Check sequence of operation.
- Replace air filter or clean reusable type filter.
- Check and inspect the ducts to make sure that there are no air leaks or discrepancies.
- Check the duct grilles and registers for proper air flow.
- Check UVC lamps, and clean and check for proper operation.
- Check smoke/fire dampers.
- Check the evaporator coil and advise if dirty or if it needs cleaning.
- If the unit you are inspecting has a belt-driven motor, check the blower belt wear, tension, and adjust as needed.
- Check bearings and lubricate blower motor if needed.
- Check voltages to the unit and check the voltage in the control box.
- Check the amperage of the blower motor and compare reading against the data information plate that is attached to the unit.
- Check the amperage of each electric heating element and compare against the data information plate that is attached to the unit.

- Check the total amperage draw for all of the electric heating elements and compare the reading against the data information plate that is attached to the unit.
- Check all of the heat sequencers for proper operation.
- Check electrical wiring and connections.
- Check the temperature rise.
- Check the supply temperature.
- Check the heat anticipator on the thermostat for the proper setting.
- Check fresh air damper for proper operation. Repair, replace, or adjust as necessary.
- Check the conditions of the entire air conditioning system and advise the owner of any discrepancies.

Heat Pump

- Check and adjust thermostat.
- Check and make sure indoor and outdoor units come on.
- Check sequence of operation.
- Replace air filter or clean reusable type filter.
- Check and inspect the ducts to make sure that there are no air leaks or discrepancies.
- Check the duct grilles and registers for proper air flow.
- Check smoke/fire dampers.
- Check the evaporator coil and condenser coil and advise if dirty or if it needs cleaning.
- Check UVC lamps, and clean and check for proper operation.
- Check the condensate drain system and secondary drain pan then advise of any discrepancies.
- Check condensate drain pan float switches for proper operation and installation.
- Check expansion valve and coil temperatures.
- If the unit you are inspecting has a belt-driven motor, check the blower belt wear, tension, and adjust or replace as needed.
- Check bearings and lubricate blower motor if needed.
- Check electrical wiring and connections.
- Check voltages to the unit and check the voltage in the control box.
- Check the blower motor amperage draw and compare the reading against the data information plate that is attached to the unit.
- Check the amperage of each electric heating element and compare against the data information plate that is attached to the unit.
- Check the total amperage draw for all of the electric heating elements and compare the reading against the data information plate that is attached to the unit.
- Check all of the heat sequencers for proper operation.

- Check the condenser fan motor bearings and lubricate if needed.
- Check the condenser motor amperage draw and compare the reading against the data information plate that is attached to the unit.
- Check compressor contactor, capacitor, and any other controls in the condenser unit.
- Check voltage and amperage draw on the compressor with a clamp-on multimeter draw and compare the reading against the data information plate that is attached to the unit.
- Check the crankcase heater on the compressor if one is installed.
- Check the defrost controls for proper operation.
- Check reversing valve for proper operation.
- If the unit you are inspecting has a start capacitor and potential relay, check it.
- If the unit you are inspecting has pressure switches, check for proper operation.
- Check all controls in control box.
- Install refrigerant gauges and check operating pressures, superheat, and subcooling.
- Check refrigerant levels and advise the owner if adjustment or repairs are necessary.
- Check fresh air damper for proper operation. Repair, replace, or adjust as necessary.
- Visually inspect exposed ducts and external piping for insulation and vapor barrier for integrity. Correct any problems found.
- Check integrity of all panels on HVAC unit. Replace missing fasteners as needed to ensure proper integrity and fit of HVAC equipment.
- Check the conditions of the entire heat pump system and advise the owner of any discrepancies.

Gas Furnace

- Check and adjust the thermostat.
- Check the heat anticipator.
- Check and see if the gas furnace comes on.
- Check sequence of operation.
- Replace air filter or clean reusable type filter.
- Check condensate drainage system and secondary drain pan and advise of any discrepancies.
- Check and inspect the air ducts to make sure that there are no air leaks or discrepancies.
- Check the duct grilles and registers for proper air flow.
- Check smoke/fire damper.
- If the gas furnace you are inspecting has a belt-driven blower motor, check the blower belt wear, tension, and adjust as needed.
- Check bearings and lubricate blower motor if needed.
- Check voltages to the unit and check the voltage in the control box.

- Check the amperage of the blower motor and compare reading against the data information plate that is attached to the unit.
- Check UVC lamps, and clean and check for proper operation.
- Check the flue for rust and corrosion and advise of discrepancies.
- Check, clean, and adjust pilot if needed.
- Check electronic spark ignition control for proper operation.
- Check all wiring and connections to controls and electrical connections.
- Check incoming gas pressures and leaving gas pressures from gas-valve manifold.
- Check burners to see if they need cleaning and advise.
- Check and adjust burners for fuel efficiency.
- Check heat exchanger for cracks, soot, and rust.
- Check the heat exchanger for cracks when the furnace is hot.
- Check blower motor and induce draft motor amperage and compare reading against the data information plate that is attached to the unit.
- Check fan controls for proper operation.
- Test safety shutoff response.
- Check the conditions of the entire gas furnace system and advise the owner of any discrepancies.
- Check fresh air damper for proper operation. Repair, replace, or adjust as necessary.
- Check integrity of all panels on furnace unit. Replace missing fasteners as needed to ensure proper integrity and fit of furnace equipment.
- Check the CO_2 monitor and the smoke alarm for proper operation; replace batteries yearly.

Dual Fuel or Heat Pump Furnace

- Check and adjust the thermostat.
- Check sequence of operation.
- Check to make sure indoor and outdoor unit come on, includes the furnace.
- Replace air filter or clean reusable type filter.
- Check and inspect the ducts to make sure that there are no air leaks or discrepancies.
- Check the duct grilles and registers for proper air flow.
- Check smoke/fire damper.
- Check the evaporator coil and advise if dirty or if it needs cleaning.
- Check the condensate drain system and secondary drain pan then advise of any discrepancies.
- Check condensate drain pan float switches for proper operation and installation.
- Check expansion valve and coil temperatures.

- If the unit you are inspecting has a belt-driven motor, check the blower belt wear, tension, and adjust or replace as needed.
- Check bearings and lubricate blower motor if needed.
- Check electrical wiring and connections.
- Check UVC lamps, and clean and check for proper operation.
- Check voltages to the unit and check the voltage in the control box.
- Check the blower motor amperage draw and compare the reading against the data information plate that is attached to the unit.
- Check the amperage of each electric heating element and compare against the data information plate that is attached to the unit.
- Check all of the heat sequencers for proper operation.
- Check the condenser fan motor bearings and lubricate if needed.
- Check the condenser motor amperage draw and compare the reading against the data information plate that is attached to the unit.
- Check compressor contactor, capacitor, and any other controls in the condenser unit.
- Check voltage and amperage draw on the compressor with a clamp-on multimeter draw and compare the reading against the data information plate that is attached to the unit.
- Check the crankcase heater on the compressor if one is installed.
- Check the defrost controls for proper operation.
- Check reversing valve for proper operation.
- If the unit you are inspecting has a start capacitor and potential relay, check it.
- If the unit you are inspecting has pressure switches, check for proper operation.
- Check the flue for rust and corrosion and advise of discrepancies.
- Check the flue for proper operation.
- Check, clean, and adjust pilot if needed.
- Check electronic spark ignition control for proper operation.
- Check and adjust burners for fuel efficiency.
- Check heat exchanger for cracks, soot, and rust.
- Check the heat exchanger for cracks when the furnace is hot.
- Check blower motor and induce draft motor amperage and compare reading against the data information plate that is attached to the unit.
- Check incoming gas pressures and leaving gas pressures from gas-valve manifold.
- Check fan controls for proper operation.
- Test safety shutoff response.
- If the unit you are inspecting has a pressure switch, check the cut-out settings.

- Install refrigerant gauges and check operating pressures, superheat, and subcooling.
- Check refrigerant levels and advise the owner if adjustment or repairs are necessary.
- Check integrity of all panels on HVAC unit. Replace missing fasteners as needed to ensure proper integrity and fit of HVAC equipment.
- Check fresh air damper for proper operation. Repair, replace, or adjust as necessary.
- Visually inspect exposed ducts and external piping for insulation and vapor barrier for integrity. Correct any problems found.
- Check the conditions of the entire system and advise the owner of any discrepancies.
- Check the CO_2 monitor and the smoke alarm for proper operation, replace batteries yearly.

Water Source Air Conditioning and Heating

- Check and adjust the thermostat.
- Check to make sure indoor and outdoor unit come on.
- Check sequence of operation.
- Replace air filter or clean reusable type filter.
- Check and inspect the ducts to make sure that there are no air leaks or discrepancies.
- Check the duct grilles and registers for proper air flow.
- Check smoke/fire damper.
- Check the evaporator coil and advise if dirty or if it needs cleaning.
- Check UVC lamps, and clean and check for proper operation.
- Check metering device and coil temperatures.
- If the unit you are inspecting has a belt-driven blower motor, check the blower belt wear, tension, and adjust as needed.
- Check bearings and lubricate blower motor if needed.
- Check the condensate drain system and secondary drain pan then advise of any discrepancies.
- Check condensate drain pan float switches for proper operation and installation.
- Check the condenser coil to determine if it needs cleaning.
- Check water source to air HVAC equipment.
- Check water source for cleanliness and proper operation.
- Inspect all wiring and connections to controls and electrical connections.
- Check voltage and amperage draw on all motors with a clamp-on multimeter.
- Check compressor contactor, capacitor, and any other controls in the condenser unit.
- Check the compressor and the amperage draw and compare the reading against the data information plate that is attached to the unit.

- Check reversing valve for proper operation.
- If the unit you are inspecting has a start capacitor and potential relay, check it.
- If the unit you are inspecting has a pressure switch, check the cut-out settings.
- Install refrigerant gauges and check operating pressures, superheat, and subcooling.
- Check refrigerant levels and advise the owner if adjustment or repairs are necessary.
- Check fresh air damper for proper operation. Repair, replace, or adjust as necessary.
- Check the conditions of the entire air conditioning system and advise the owner of any discrepancies.
- Check integrity of all panels on HVAC unit. Replace missing fasteners as needed to ensure proper integrity and fit of HVAC equipment.

Testing, Adjusting, and Balance

HVAC system testing, adjusting, and balancing is the process checking and adjusting all environmental systems in a building to return the HVAC equipment back to the original design conditions. Test, adjust, and balance (TAB) of an HVAC system may have to be done more than once due to seasonal changes or any structural changes in the property. The effectiveness of any HVAC system is completely dependent on the amount of airflow through it. The airflow through a HVAC system is dependent on an adequately sized evaporator (air handler) fan blower; a properly sized, balanced, and designed air duct system; clean evaporator coil(s); and non-restrictive air filters in the air handler and clean condenser coil(s). Buildings that use water source equipment and cooling tower will also need to be tested, adjusted, and balanced.

Air Duct Distribution Preventive Maintenance

Commercial forced air conditioning and/or commercial forced heating systems efficiency is affected by the following:

- Air duct leakage
- Air conditioning and/or heating system airflow
- Blower operation
- Air ducts balanced between the supply air and the return air
- Condition of the air duct system and insulation

If the air ducts (Figure 12-1) are not inspected, adjusted, repaired, or replaced first, the air conditioner refrigerant charge cannot be checked properly or recharged correctly with refrigerant. You will have poor superheat or poor subcooling temperatures. As for the heating systems, if the air ducts are not inspected, adjusted, repaired, or replaced, the heating system might not function properly or it might overheat and the system will shut down. HVAC contractors and service technicians servicing HVAC equipment must also determine if the air duct distribution system is adequate for the HVAC equipment. Undersized ducts will decrease efficiency and increase energy costs.

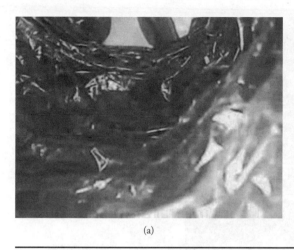

(a)   (b)

Figure 12-1   (a) A view from inside of the flexible duct with a video camera—with a camera you can determine if the duct is in need of repairs or blocked with debris; (b) an internal view of a fiber glass duct with a video camera—with a camera you can determine condition and cleanness of the duct.

Figure 12-2
Improper installation
of flexible duct.

Figure 12-3
What do you see
wrong in this figure?

Inspecting Air Ducts

Before starting your quest through the attic or ceiling, you will need to prepare the duct system. You must bring a trouble light and a flashlight, to guide you through the darkness; bring a mirror too, this will help you see and locate leaks under the ducts. Put on safety glasses and wear a dust mask, this will prevent you from breathing the dust in an enclosed area when you begin your inspection. One of the simplest ways of finding air duct leaks is to look for the obvious (Figures 12-2 to 12-5) and by feeling with your hand for air leaking from the supply ducts, while the ducts are pressurized by running the air-handler blower fan. Closing the dampers on supply registers temporarily or by blocking the register with plastic, or with any object that won't be blown off by the registers airflow. This will increase the duct pressure within the duct and makes looking for air leaks easier. When inspecting air ducts look for the obvious.

Another way to locate air leaks is to insert a trouble light with a 100-W bulb inside the duct through a register. Look for light emanating from the exterior of the duct joints and seams. Some technicians have in their arsenal of tools a video inspection camera; this is used to inspect the duct system from the inside.

Checking the Cooling System for Proper Refrigerant Charge

Only an EPA-certified technician can check the refrigerant charge in the air conditioning unit. For a non-thermostatic expansion valve (TXV) systems, the technician will have to check the refrigerant charge for the proper superheat according to the manufacturer's specifications as listed in the service literature. For TXV systems, the technician will have to check the refrigerant charge for the proper subcooling according to the manufacturer's specifications as listed in the service literature. Most manufacturers place the refrigerant charging information on the inside of the control panel in the condensing unit. Always follow the manufacturer's recommendations for checking or charging the air conditioning equipment.

Filtration

Air filtration removes unwanted air particles (dust) from the air stream in an air conditioning system within a building. There are two methods of air filtration to purify or filter the air: mechanical filtration and electrostatic filtration. Mechanical filtration consists of air passing through a filter medium that captures the particles from the air passing through the air conditioning equipment and reentering the conditioned space. As mechanical filters load with particles over time their collection efficiency and pressure drop (air resistance) increase. Over time, the increased pressure drop will inhibit the air flow causing the air conditioning equipment to work harder, losing efficiency and increasing energy costs. As the pressure drop increases across filter media, it is an indicator that the filters will have to be replaced.

Check the service literature for the model unit being serviced for the correct type of filter needed. The MERV rating for filters is listed in Table 12-1.

The HVAC technician must also be aware of the different types of filters available on the market today. If the wrong type of filter is used then it can decrease the efficiency of the equipment and increase energy usage.

Filtration Maintenance

Filter maintenance is crucial to keeping HVAC ductwork and HVAC equipment clean. If dust and dirt accumulates in the HVAC ducts and in the air-handler section, mold and bacteria may form, causing indoor air quality issues (Figure 12-6). For this reason, it is crucial to establish the appropriate filter change-out frequency as provided

Typical Air Filter Type	Disposable Panel Filters, Fiberglass and Synthetic Filters, Permanent Self-Cleaning Filters, Electrostatic Filters, Washable Metal Foam Filters	Pleated Air Filters, Extended Surface Air Filters, Media Panel Filters	Non-Supported Bag Filters, Rigid Box Filters, Rigid Cell/ Cartridge Filters	Non-Supported Bag Filters, Rigid Box Filters, Rigid Cell/ Cartridge Filters MERV 13 Pleats	HEPA Filters, ULPA Filters, SULPA Filters
MERV Std. 52.2	**1–4**	**5–8**	**9–12**	**13–16**	**17–20**
Average dust spot efficiency	<20%	<20–35%	40–75%	80–95%+	99.97% 99.99% 99.999%
Average arrestance ASHRAE Std. 52.1	60–80%	80–95%	>95–98%	98–99%	N/A
Particle size ranges	>10.0 μm	3.0–10.0 μm	1.0–3.0 μm	0.30–1.0 μm	<0.30 μm
Typical air filter applications	Residential, light commercial, equipment protection	Industrial workplace, commercial, paint booths	Industrial Workplace, high-end commercial buildings	Smoke removal, general surgery, hospitals, and health care	Clean rooms, High-risk surgery, hazardous materials, pharmaceutical manufacturing
Types of things filter will trap	Pollen, dust mites, standing dust, spray paint dust, carpet fibers	Mold spores, hair spray, fabric protector, cement, dust	Humidifier dust, lead dust, auto emissions, milled flour	Bacteria, most tobacco, smoke, proplet nuceli (sneeze)	Viruses, radon progeny, carbon dust

TABLE 12-1   ASHRAE Standard 52.2 Filter Rating Chart

in the service literature from the manufacturers of the HVAC equipment and the filter manufacturer/supplier. Filters should be replaced when

- the filter becomes wet,
- microbial growth becomes visible on the filter media,
- the filter collapses into the air handler, or
- the filter becomes damaged allowing the return air to bypass the filter media.

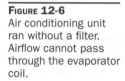

Figure 12-6
Air conditioning unit ran without a filter. Airflow cannot pass through the evaporator coil.

When installing a new or cleaned filter, pay close attention when reinstalling the filter into the filter frame or rack, you want to avoid air bypassing the filter media. Make sure that all of the air will pass through the filter media. Listed below are some more tips:

- Check filters for holes, rips, and damage; replace the filter if any of these are present.
- Check and make sure the filter media is sealed in the filter frame.
- Install the filter according to the airflow direction indicated on the filter frame.
- Make sure the filter fasteners are correctly installed and are in place.
- To avoid collapse of the filter frames, make sure the bank of filter frames is rigid and well reinforced.
- Seal all openings that will allow bypass air to pass between the filter frames or filter banks.
- When replacing and reinstalling the new filters, pay attention to the condition of the filter frame seals and gaskets; replace as necessary.
- Check and make sure the filters match the filter frame size, if they do not, replace the filter(s) with the correct size; this will prevent air bypass.

When Do Filters Get Replaced?

When do you replace the filter(s)? This will depend on the type of filter media used. Do not rely on a visual inspection only, since medium and high efficiency filters may look dirty, they might not have reached their optimum efficiency levels yet. Some commercial HVAC units may have a differential pressure measurement sensor installed across the filter bank to identify the appropriate time to replace the filters. This pressure drop switch device is wired up to an alarm, signal light, or to the building automation network, alerting operators it's time to replace the filters. When filters load up with dust particles, they become more efficient. But, the pressure drop will increase causing reduced airflow in the HVAC unit. When the filters have reached their excess limit, they will become deformed and the filters will blow out causing the unfiltered dust particles to enter into the coils and air ducts. When this condition happens, the indoor coils and air ducts will become dirty, causing an additional maintenance cost to clean the coils and air ducts.

Electronic Air Cleaners

An active electrostatic filter (electronic filter) consists of a high-voltage charge between plates, either a positive or negative charge applied. The dust particles will pass over the plates and will be electrostatically withdrawn from the passing air and captured on the charged plates. This type of filtration system needs constant maintenance performed on it. If maintenance is not performed constantly the efficiency rating will begin to decrease due to rapid build-up on the collection plate.

Electronic Air Cleaner Maintenance

At the beginning of the cooling or heating season, the collecting cell and pre-filter assembly should be cleaned. The maintenance on the electronic air cleaner should always be performed according to the manufacturer's instructions as stated in the service literature. However, the environment dictates the frequency for cleaning the electronic air cleaner. The technician should follow the safety precautions listed in the service literature before cleaning and testing the electronic air cleaner.

The following maintenance procedure is just a general procedure for cleaning an electronic air cleaner. First, turn off the electricity to the electronic air cleaner. If the model air cleaner you are servicing is attached to a cooling system, turn the electricity off to that unit also. Next, gain access to the filters and collecting cell assembly and remove them from the electronic air cleaner cabinet (Figure 12-7). Now, thoroughly vacuum the inside cabinet of the electronic air cleaner. If the blower assembly needs cleaning, vacuum the dust off the blower assembly and from the blower wheel (Figure 12-8). Some electronic air cleaners contain UVC lights; the light bulb(s) will have to be wiped clean, removing dust accumulation. Take the pre-filter and vacuum it from the entering airside,

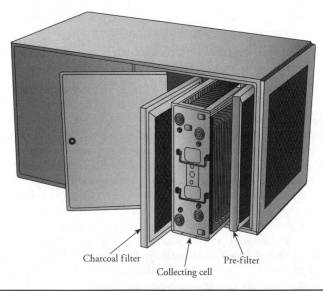

Charcoal filter

Collecting cell

Pre-filter

FIGURE 12-7 Gain access and remove to the filters and collecting cell assembly.

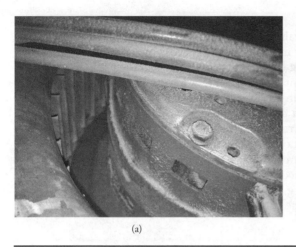 

(a)   (b)

FIGURE 12-8   (a) Inspect the blower motor for dirt accumulation and vacuum the openings in the motor body; (b) make sure the dirt is removed from the blower wheel vanes.

this will make easier to clean. If vacuuming does not get the pre-filter clean, then it will have to be washed. Locate a suitable container large enough to soak the collecting cell(s) and pre-filter (Figure 12-9). Place ¾ of a cup of dishwasher detergent per collecting cell into the container; place the collecting cell(s) and pre-filter into the container. Add enough hot water to cover the collecting cell(s) and soak them for 15 to 20 minutes. Agitate the cell(s) every few minutes. This action will aid to loosen the dirt from the collecting cell(s) Make sure the detergent is totally dissolved. When it is time to remove the collecting cell(s) and filter, rinse them off with clear clean water. Discard the dirty water from the container in an approved manner and rinse the container with clear clean water to remove any sediment. Next, fill the container with hot clear clean water and place

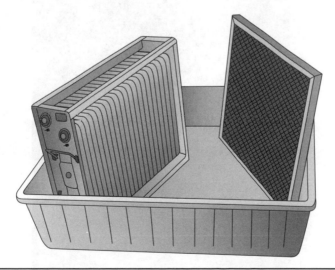

FIGURE 12-9   Place filter and collecting cell assemble in a pan large enough to be able to soak the components in dishwasher detergent.

the collecting cell(s) in the container; soak for another 15 to 20 minutes. When the time is up, remove the collecting cell(s) and let the water drain from the cell(s). If the water draining feels slippery, repeat the procedure again. Allow the pre-filters to dry thoroughly for about 2 hours before re-installing. Inspect the collecting cell(s) and make sure that there are no broken wires, broken or bent ionizer springs, bent collector plates, deformed ionizer grid, or cracked or broken insulators. Repair or replace any damaged collecting cell. Observing the airflow arrows, reinstall the collecting cell(s) and the pre-filter back into the electronic air cleaner cabinet. The charcoal filter loses its odor removal efficiency within 6 months. The charcoal filter cannot be washed, it must be replaced. To ensure the collecting cell(s) are completely dry, restore the electricity to the electronic air cleaner, run only the blower fan for about ½ hour. This action will ensure the collecting cell(s) will be dried out. After ½ hour, turn on the electronic air cleaner, if arcing (loud cracking noise) occurs, it will last for a few minutes, and arcing will not damage the collecting cell.

Condensate Management and Drain Systems

Condensate management systems are used to monitor, detect, and arrest condensate water overflow in HVAC equipment. The condensate detection method can be as simple as a float switch only or a complete management system including an alarm connected to the HVAC equipment. When the presence of any condensate water overflow is detected (Figures 12-10 and 12-11), the condensate management system automatically turns off the

FIGURE 12-10
Condensate water overflowing the pan.

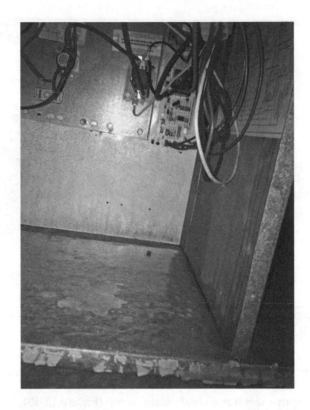

FIGURE 12-11
Condensate spillage extended past the condensate pan. This needs to be cleaned up before returning the air conditioning unit on.

HVAC unit. By turning off the HVAC unit, the condensate management system will protect commercial buildings from potential water damage to ceilings, walls, woodwork and flooring, carpets, furniture, office equipment, and other commercial property.

To turn the HVAC unit back on, the water in the condensate pan and the water in the secondary condensate pan will have to be removed (Figure 12-10). This is accomplished by using a wet/dry vacuum to remove the condensate water from the pans. When the water is completely removed, the float switches (Figure 12-12) will return to normal operation and the HVAC equipment now operates. The condensate drain line will also need to be cleared of any blockages also.

Condensate Pump Maintenance

A condensate pump is used when a gravity drain is impractical or impossible to install in a commercial building. The plumbing from the HVAC equipment to the condensate pump reservoir should be installed with traps and float switches as if the condensate water was draining like a typical gravity drain. When the condensate water leaves from the condensate drain pan the water will drain into the condensate pump reservoir tank. As the water rises, a float switch turns on the pump motor and the water is pumped from the reservoir tank, usually through clear plastic tubing, to a drain or outside of the building.

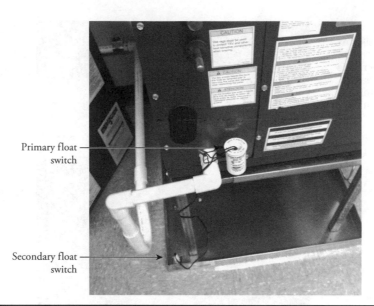

Primary float switch

Secondary float switch

FIGURE 12-12 Emergency drain pan and secondary drain pan with emergency condensate float system.

Should the condensate pump fail or the discharge line become blocked, a second switch will open a low-voltage circuit that shuts off the HVAC equipment and even can be wired to set off an alarm if needed.

If the HVAC equipment has a condensate pump (Figure 12-13), it too would need to be inspected and cleaned. Make sure the condensate pump is unplugged or turn off the circuit breaker before attempting servicing. Next, remove the condensate pump from the unit and drain the water from the reservoir. The condensate pump pan can be separated from the reservoir for cleaning. Thoroughly clean the reservoir of any grime and dirt. Clean the float and make sure it moves freely. If you are uncertain on how to service the condensate pump, follow the instructions in the service literature from the condensate pump manufacturer.

When you have completed the maintenance on the condensate pan(s), check with local codes and laws or equipment manufacturers, if you can place approved chemicals or tablets in the condensate pan to help prevent the buildup of slime or algae.

Fire and Smoke Dampers

A fire damper (Figure 12-3) closes once the air duct temperature reaches a high enough level to melt a fusible link. Its primary function is to prevent the passage of flame from one side of a fire-rated separation to the other side. The fusible link in a fire damper is rated for 165°F up to 286°F. There are two types of applications for fire dampers: static and dynamic. Static fire dampers can only be applied in HVAC systems that are designed to shut down in the event of a fire.

A smoke damper closes upon detection of smoke and is operated by either electricity or by a pneumatic actuator. This type of damper is designed to prevent the circulation of air

A condensate pump located in a ceiling in an office building.

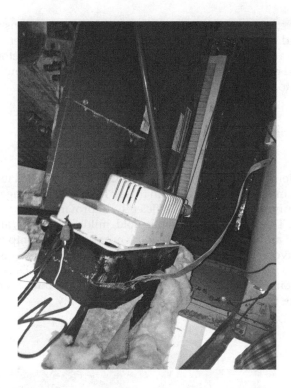

and smoke through a duct or a ventilation opening. Duct detectors are installed in the air ducts, sampling the air for smoke or fumes. Smoke detectors are connected to the building's fire alarm system. If smoke is detected, the dampers will close, stopping the fans and blowers, and triggering an audible alarm. These actions provide early detection allowing the occupants time to leave the building and the fire alarm system will contact the fire department. The duct detectors are very sensitive to dirt buildup, this can trigger false alarms. Preventive maintenance information is specific to each manufacturer and should be strictly adhered to. Following are some suggestions for maintenance procedures:

- Every 6 months inspect the smoke detectors to confirm they are in a normal operating state.
- Annually test the units to initiate an alarm, ensuring the smoke detectors produce the intended response.
- On new installations, check within 1 year after the installation and every other year thereafter to ensure the smoke detectors are within their listed and marked sensitivity range.
- Annually test the air to ensure the smoke detectors are sampling the airflow in the ducts.

Air conditioning air handlers are hard to reach most of the time because they are located in the ceilings. Remote test or reset switches that are easy to see and access should be installed close to the air handler.

Combination fire/smoke dampers are used in HVAC penetrations where a wall, floor, or ceiling is required to have both a fire damper and a smoke damper. They close upon the detection of heat (via duct temperature) or smoke (via a smoke detector) and seal the opening. Preventive maintenance information is specific to each manufacturer and should be strictly adhered to.

UVC Lights

Some air handlers or package units have a UVC lamp(s) installed in it (Figure 12-14). The UVC lamp reduces microbial growth within the air handler or package unit, allowing the air conditioning equipment to run more efficiently, improving indoor air quality within the occupied space. The UVC lamp prevents mold, mildew, bacteria, and viruses from collecting inside the air handler and attached plenum area. This lamp must be inspected every time preventive maintenance is done on the air conditioning equipment. With the electricity turn off to the UVC lamp, wipe the dust from the bulb. Depending on which type of UVC lamp is installed, you might have to replace the UVC bulb every 1 to 2 years. Check the service literature that came with the UVC lamp assembly.

FIGURE 12-14
Some air handlers have UVC lamps installed. Check UVC lamp operation. Replace UVC bulbs when necessary or recommended by the UVC manufacturer. See UVC lamp service literature for proper instructions.

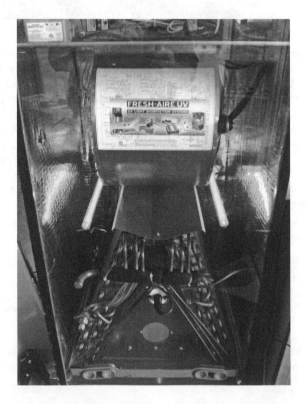

SAFETY NOTE *Human exposure to UVC light may result in temporary eye (cornea) damage and skin (sunburn) damage. While these effects are mostly temporary, they can still be very painful. Service technicians servicing air conditioning equipment that has UVC lamps installed in it should be aware of hazards and take proper measures for self-protection. The first step when servicing an air conditioning unit, is to turn off the electricity to the air conditioning unit or UVC lamp assembly. If for any reasons the technician needs to directly view the lamps, follow these guidelines:*

- *When temporarily viewing lamps from 4 ft or more, you will need to wear eye protection with polycarbonate lens (rated for UV) and the lenses must wrap around the face to fully protect the eyes (goggle-style eyewear).*

- *The most critical, when directly exposed to UVC less than 4 ft, it is necessary to wear full face and skin protection. This viewing is not suggested unless the service technician has been properly trained.*

Operational Checks on Air Conditioning Equipment

After the cleaning and mechanical inspections are completed on the air conditioning and heating equipment, the operation of the air conditioning and heating equipment must be checked to determine if its operating according to the manufacturer's specifications. The operating checks and adjustments include:

- Input voltage checks to the equipment
- Current checks on the equipment
- Cooling system checks on the equipment
- Heating system checks on the equipment's electric resistive heaters
- Heating system checks on the heat pump

Sequence of Operation   Information on the correct sequence of operation for the model unit you are checking is found in the service literature.

Input Voltages and Current to the Air Conditioning and Heating Equipment

- Test the voltage input and the current to the equipment while it is operating with a clamp-on ammeter.
- Compare the readings against the name-data tag on the unit or with the service literature.

Water Source Air Conditioning and Heating Preventive Maintenance

To perform the preventive maintenance on a water source air conditioning unit, follow the steps listed under "Water Source Air Conditioning and Heating" in this chapter. Instead of a fan removing the heat from the refrigerant in the condenser coil, water is circulated through a "tube in a tube" type of condenser coil (Figure 12-15) to

FIGURE 12-15   A water cooled condenser coil.

remove the heat from the refrigerant. The water source can come from a surface water source such as a lake, river, or stream. The water can also come from a groundwater source such as a well or aquifer. Depending on the region, surface water will have seasonal variations and can carry silt and debris that can cause a fouling if the water is not pre-filtered before entering the condenser coil. Groundwater does not have seasonal variations. But, depending on the geology of the region, the water source can have high levels of dissolved minerals that contribute to scale formation on the inside of the condenser coil. The scale formation on the inside of the condenser coil acts as an insulator which prevents heat transfer from the refrigerant to the water. Some areas use a water tower that reuses the water by recirculating it from the tower to the cooling equipment and back to the tower again. Whichever type of water source is used, in due time the condenser coil will have to be cleaned. To clean the condenser coil follow the instructions listed in the service literature.

Water-Cooling Towers

This chapter will only cover the basics needed to perform the preventive maintenance on water-cooling towers. The maintenance procedure on a cooling tower should be done by a professional service technician trained in servicing the cooling tower.

Expert

Hard

Moderate

Easy

Call a certified HVAC service Company

Service technician

Business Owner

Home Owner

Following are symbols that are used in this chapter to define who should perform the preventive maintenance on the residential refrigeration appliances.

Safety First

Any person who cannot use basic tools or follow written instructions should *not* attempt to install, maintain, or repair any water-cooling towers. Any improper installation, preventive maintenance, or repairs could create a risk of personal injury or property damage. If you do not fully understand the preventive maintenance procedures in this chapter, or if you doubt your ability to complete the task on the cooling tower, please call your HVAC service manager.

Water-Cooling Towers in General

Much of the preventive maintenance information in this chapter covers water-cooling towers in general, rather than specific models, in order to present a broad overview of preventive maintenance techniques. The illustrations that are used in this chapter are for demonstration purposes only, to clarify the description of how to perform preventive maintenance on the cooling tower. They in no way reflect on a particular brand's reliability.

Water-Cooling Towers

Water-cooling towers (Figure 13-1) are heat-transfer units that use water and air to transfer heat from the air conditioning systems to the outdoor environment. Cooling towers are used with water-source air conditioning equipment and chillers, and are usually located on rooftops or on outdoor sites on the property. Water-cooling towers fall under two categories: open circuit and closed circuit. In open-circuit systems, the recirculating water travels through the water-source air conditioning equipment or chiller and returns to the tower, distributed across the heat-transfer medium (Figure 13-2), and are exposed to the atmosphere. In closed-circuit systems, the recirculating water, mixed with glycol, circulates through the water-source air conditioning equipment or chiller, back to the tower, through a coil located within the tower structure, while the cooling tower water recirculates the water in the cooling tower only, removing the heat from the coil. The water circulating within the coil never comes in contact with the atmosphere. In both cooling tower categories, make-up water must be introduced to replace the water that evaporated from the tower. To circulate the water throughout the building, an external pump and motor is used (Figure 13-3). The water must be chemically treated to prevent scale formation (Figure 13-4). Scaling causes clogging in the pipes, affecting the flow of air or water, and reducing the efficiency of the cooling tower. Corrosion and oxidation are other problems that occur in cooling towers. Corrosion is caused in metal over time due to environment conditions, and oxidation will eat the remaining exposed metal in the cooling tower.

(a)   (b)

FIGURE 13-1 (a) A small 20-ton cooling tower; (b) a large cooling tower used in conjunction with a chiller system.

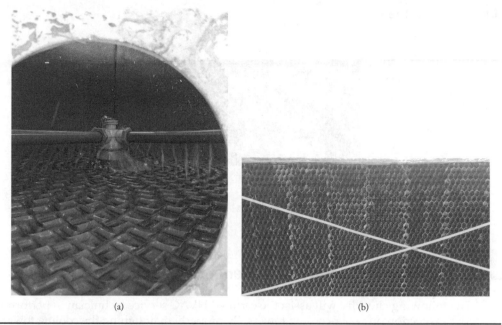

FIGURE 13-2 (*a*) Water spraying downward across the heat-transfer medium; (*b*) the water is broken up into smaller particles for easier heat rejection.

FIGURE 13-3 (*a*) Small water recirculating pump for a small cooling tower; (*b*) larger water recirculating pumps used for large rooftop cooling towers.

FIGURE 13-4
FIGURE 13-4
Chemicals are used to treat the recirculating water in cooling towers.

Water Cooling Tower Preventive Maintenance Checklist

The following checklist will assist the certified HVAC service technician in performing a proper inspection on water-cooling towers. By properly maintaining the cooling tower, the efficiency of the equipment will be operating at or above the 90% range of the rated BTU capacity with a reduction in energy use. Always check the manufacturer's literature that comes with the cooling tower if there are additional preventive maintenance checks that are not listed on this checklist.

Water-Cooling Towers

- Check operation sequence
- Perform a visual inspection
- Check the fan motor operation
- Check fan blade(s)
- Check screens
- Check the suction screen
- Check the make-up water float and switch
- Check for excess vibrations
- Check the cooling-tower structure
- Check belts and pulleys
- Test the water
- Check lubrication on bearings
- Check motor supports and brackets
- Check motor alignment
- Check drift eliminators, louvers, and fill
- Check sump and spray nozzles for clogging

- Check gear box
- Inspect wiring and connections
- Check motor starting contacts for wear and proper operation
- Check sump heater
- Check pumps
- Check for water leaks

Maintenance Schedule for Water-Cooling Towers

The following maintenance schedule will assist building owners and service technicians in the preventive maintenance frequency on the cooling tower:

Daily

- Turn on the water cooling tower.
- Complete overall visual inspection of the cooling tower to be sure all equipment are operating and safety systems are in place.

Weekly

- Check the condition of the fan motor and blade through temperature or vibration analysis and compare the findings to the baseline values.
- Physically clean the screens including the suction screen of all debris.
- Operate the make-up water float switch manually to ensure proper operation.
- Check for excessive vibration in motor(s), fan(s), and pump(s).
- Check the cooling tower structure for loose fill, connections, leaks, and the like.
- Check all belts and pulleys, adjust or replace as necessary.
- Test water samples for proper concentrations of dissolved solids, and chemistry. Adjust blowndown and chemicals as necessary. Perform this procedure for open-circuit-type cooling towers once a week and for closed-type cooling towers monthly.

Monthly

- Check lubrication for all bearings. Bearings are lubricated per manufacturer's recommendation.
- Check motor supports and fan blades for excessive wear and secure fastening.
- Check the motor alignment and motor coupling alignment, this allows for efficient torque.
- Check the drift eliminators, louvers and fill for proper positing and scale buildup.

Annually

- Inspect the water nozzles for clogging, make sure water is flowing through the nozzles.

- Clean the cooling tower. Remove all dust, scale, and algae from the tower basin, fill, and spray nozzle.

- Inspect the bearings and drive belts for wear. Adjust, repair, or replace as necessary.

- Check the condition of the fan motor and blade through temperature or vibration analysis and compare the findings to the baseline values.

14
CHAPTER

Self-Contained, Commercial Refrigerators and Freezers

This chapter will only cover the basics needed to perform the preventive maintenance on self-contained, commercial refrigerators and freezers. Each preventive maintenance procedure has been rated by a degree of complexity or simplicity. For the owner of the product, you will have to decide whether you can perform the preventive maintenance procedure or if you must call a professional to complete the preventive maintenance procedures on the product. Following are symbols that are used in this chapter to define who should perform the preventive maintenance on the self-contained, commercial refrigerators and freezers.

Expert	Call a certified HVAC service Company
Hard	Service technician
Moderate	Business Owner
Easy	Home Owner

The business owner will need hand tools and a flashlight and drop light, to gain access and to see into the self-contained, commercial refrigerator or freezer.

The chapter will not cover diagnostics, any electrical repairs, or replacement procedures of any sealed-system components. This is a specialized area, and only an EPA-certified technician can repair the sealed system on self-contained refrigerators or freezers. The certified technician has the proper training, tools, and equipment required to make the necessary repairs to the sealed system. For the new HVAC/R student, the information in this chapter will be of great importance in your studies to become a certified air conditioning/refrigeration technician. An individual who is not certified to service the sealed system could raise the risk of personal injury, as well as property damage. The uncertified individual could void the manufacturer's warranty on the product if he or she attempts entry into the sealed system.

Self-Contained Commercial Refrigerators and Freezers

A self-contained commercial refrigerator and a self-contained commercial freezer contain a compartment that holds a compressor, condenser coil, condenser fan, condensate drain pan, and controls. The food storage area within the cabinet or case holds an evaporator coil, evaporator fan(s), defrost system in freezers only, temperature and defrost control and it stores the food or frozen food product, or ice product that will be preserved until it is needed to be prepared for consumption or sold to the consumer.

These self-contained refrigeration products are in supermarkets, convenience stores, restaurants, cafes, hotels and motels, cafeterias, drug stores, taverns, clubs, flower shops, ice cream stores, fast food stores, hospitals and medical facilities, schools, and any other establishment that uses refrigeration to store and dispense food, ice, and drink products. Figures 14-1 through 14-32 are just some types and styles of the self-contained commercial refrigerators and freezers. These products are available in a high-temperature range (45°F and above), in a medium-temperature range (35°–45°F) or in a low-temperature range (below 35°F). The capacity of the self-contained commercial refrigerator or freezer will vary on usage, storage of the food or ice product, or food preparation. Also available are walk-in self-contained commercial refrigerated coolers and freezers.

FIGURE 14-1 A self-contained commercial refrigerated preparation unit. This illustration shows two refrigerators below, food preparation table, and food storage on top.

FIGURE 14-2 A self-contained four-door commercial refrigerator.

FIGURE 14-3   A self-contained commercial chest freezer with sliding see-through glass doors.

FIGURE 14-4   A self-contained commercial two-door vertical refrigerator.

FIGURE 14-5   A self-contained commercial refrigerated wine storage.

FIGURE 14-6   A self-contained commercial single-door freezer on the left and a refrigerator on the right.

FIGURE 14-7   A self-contained commercial side-by-side refrigerator. This model is also available in matching freezer.

FIGURE 14-8   A self-contained commercial vertical reach-in refrigerator with see-through glass door.

FIGURE 14-9   A self-contained commercial reach-in refrigerator with sliding glass doors.

FIGURE 14-10   A self-contained commercial double glass door reach-in refrigerator.

FIGURE 14-11 A self-contained commercial island display case.

FIGURE 14-12 A self-contained commercial roll-around display case. This case at the time the picture was taken, displayed meats.

FIGURE 14-13 A self-contained commercial double deck roll-around reach-in display case on the left. On the right is a closed-in display case, assessable through the rear. Both cases were used for seafood when this picture was taken.

FIGURE 14-14 A self-contained commercial vertical reach-in freezer with see-through glass door, for storing bagged ice.

FIGURE 14-15   A self-contained commercial three door, multi-deck, reach-in freezer.

FIGURE 14-16   A self-contained commercial three door, multi-deck reach-in refrigerator with LED lights.

FIGURE 14-17   A self-contained commercial refrigerated soda machine.

FIGURE 14-18   A self-contained commercial refrigerated multi-deck vertical display case with glass see-through doors.

FIGURE 14-19   A self-contained commercial reach-in display case used for beverages.

FIGURE 14-20   A self-contained commercial ice on beverage design dispenser with chilled water.

FIGURE 14-21   A self-contained commercial vertical multi-deck refrigerator with sliding see-through doors.

FIGURE 14-22   A self-contained commercial ice cream dispenser machine.

FIGURE 14-23 A self-contained commercial single-door refrigerator.

FIGURE 14-24 A self-contained chest freezer with curved sliding doors.

FIGURE 14-25 A self-contained commercial refrigerated open floral case.

FIGURE 14-26 A self-contained commercial roll-around display case. The case is used for seafood at the time the picture was taken.

Figure 14-27   A self-contained commercial refrigerated lobster tank.

Figure 14-28   A self-contained commercial roll-around display case.

Figure 14-29   A self-contained refrigerated beverage dispenser with chilled water.

Figure 14-30   A self-contained commercial chest freezer for storing dry ice.

FIGURE 14-31 A self-contained commercial refrigerated reach-in case with a hot reach-in display on top.

FIGURE 14-32 A self-contained commercial refrigerated blast chiller.

Principles of Operation

The refrigeration cycle (Figure 14-33) in a self-contained commercial refrigerators and freezers is similar to the refrigeration process described in Chapter 11. Following is a summary of the refrigeration cycle.

Step 1. The refrigerant circulating within the evaporator coil (located near the food product) absorbs the heat from the food product being cooled. The refrigerant within the evaporator coil will change state from a liquid to a vapor and travel to the compressor.

Step 2. The compressor compresses the refrigerant vapor, from a low-pressure and low-temperature vapor, into a high-pressure and high-temperature vapor. The refrigerant leaving the compressor will begin to enter the condenser coil.

Step 3. The high-pressure and high-temperature vapor refrigerant enters the condenser coil where the refrigerant will give up the heat from the compressor and the heat is absorbed from the food product. The heat will be dispersed into the atmosphere. When the refrigerant leaves the condenser coil it will be in the high-pressure liquid form.

Step 4. The liquid refrigerant leaves the condenser coil; it is a high-pressure lower-temperature refrigerant. The temperature of the refrigerant leaving the condenser

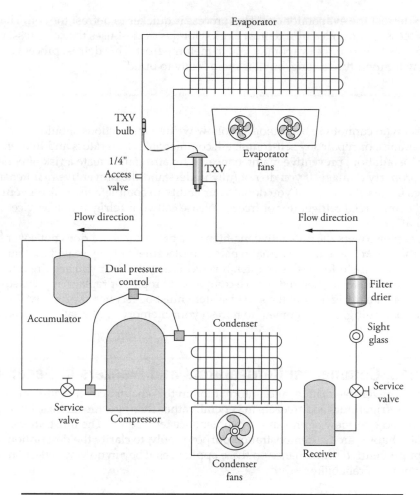

FIGURE 14-33   A typical refrigerant circuit used in a self-contained
refrigerator or freezer.

coil is cooler than the entering refrigerant temperature. Next, the refrigerant will
flow through a metering device [thermostatic expansion valve (TXV)] or a capillary tube which will reduce the pressure and temperature of the refrigerant when
it enters the evaporator coil.

Self-Contained Commercial Refrigerators and Freezers Defrost Methods

Off-cycle defrost, electric heater element defrost, and hot-gas defrost are the common
techniques used to defrost the evaporator coil. The off-cycle defrost method defrosts the
evaporator coil by cycling the evaporator fan and by stopping the flow of refrigerant. This
method is used when the temperature is 2° above freezing temperatures. The moving air in
the case warms up and melts the ice from the evaporator coil. Electric resistance heating
defrost defrosts the evaporator coil by applying heat from the electric heaters, turning off

the fan, to defrost the evaporator coil. This process is quicker to defrost the coil. The electric defrost process is controlled by a defrost clock. Hot-gas defrost uses the compressor hot refrigerant gas to warm the evaporator coil to melt the frost. This defrost process can be used in any temperature setting, but it is more costly to build.

Safety First

Any person who cannot use basic tools or follow written instructions should *not* attempt to install, maintain, or repair any self-contained, commercial refrigerators and freezers. Any improper installation, preventive maintenance, or repairs could create a risk of personal injury or property damage. If you do not fully understand the preventive maintenance procedures in this chapter, or if you doubt your ability to complete the task on your self-contained commercial refrigerator or freezer, please call your refrigeration service professional.

This chapter covers the preventive maintenance procedures but does not cover how to diagnose the sealed system. The actual repair or replacement of any sealed-system component is not included in this chapter. It is recommended that you acquire refrigerant certification (or call an authorized service company) to repair or replace any sealed-system component, as the refrigerant in the sealed system must be recovered properly.

Before continuing, take a moment to refresh your memory on the safety procedures in Chapter 1.

Self-Contained, Commercial Refrigerators and Freezers in General

Much of the preventive maintenance information in this chapter covers self-contained, commercial refrigerators and freezers in general, rather than specific models, in order to present a broad overview of preventive maintenance techniques. The illustrations that are used in this chapter are for demonstration purposes only, to clarify the description of how to perform preventive maintenance on these appliances. They in no way reflect on a particular brand's reliability.

Preventive Maintenance Checklist

The following checklist will assist the certified HVAC/R service technician in performing a proper inspection on self-contained commercial refrigerators and freezers. By properly maintaining the refrigeration equipment, the efficiency of the equipment will be operating at or above the 90% range of the rated BTU capacity with a reduction in energy use. Always check the manufacturer's literature that comes with the self-contained commercial refrigerators or freezer if there are additional preventive maintenance checks that are not listed on this checklist.

The following checklist is for preventive maintenance provided by a service technician on a quarterly basis, for self-contained commercial refrigerators and freezers:

- Check for any damages to the refrigerator or freezer.
- Visually inspect for corrosion on the outer and inner cabinet.

- Visually inspect for corrosion of evaporator coil and condenser coil fins, copper tubing, and solder joints.
- Check for vibrations or abnormal noise from fan motors and blades, cabinet panels, and tubing.
- Check and make sure the refrigerator or freezer is level from side to side and front to rear.
- Inspect refrigerant lines, confirm they are secured, not touching or rubbing against other refrigerant lines, wires, or frame work.
- Inspect and verify evaporator and condenser fan operation.
- Check fan motor mounts. Make sure they are tight.
- Check fan blades; make sure they are secure and not rubbing or hitting against anything.
- Check all wiring and connections for tightness. Replace any damaged wiring.
- Check all lamp connections for tightness and if they are dry.
- Check for any air disturbances external to the refrigerator or freezer. Air conditioning or heat grilles or registers, fans, and doors, and the like that might affect equipment operation.
- Check for any water leaks.
- Record condenser inlet and outlet temperatures.
- Check for restrictions of the condenser air inlet and exhaust.
- Visually check for refrigerant leaks and oil leaks.
- Record case product temperature.
- Record the return and discharge air temperatures.
- Record ambient conditions surrounding the self-contained refrigerator or freezer (wet- and dry-bulb temperatures).
- Check product loading against manufacture's specifications. Advise store manager or owner if overloading occurs.
- Check proper clearances on sides and rear of self-contained refrigerator or freezer.
- Check all door switches for proper operation.
- Check all doors and lids for proper operation and proper sealing.
- Inspect that all panels, grilles, and covers are properly installed.
- Check door gaskets for proper sealing.
- Check door hinges and door closers.
- Check moisture indicator/sight glass for proper refrigerant level, if equipped.
- Check insulation on piping.
- Check anti-sweat and condensate heaters for proper operation.

The following checklist is for preventive maintenance by a service technician on a semi-annual basis, for self-contained commercial refrigerators and freezers:

- Clean all evaporator coils and evaporator fan blades.
- Clean discharge air honeycombs or grilles.
- Clean condenser coils and condenser fan blades.
- Check the waste outlet for debris.
- Clean condensate drain pan and drain line.
- Verify condensate drain line is functioning properly.
- Verify compressor amperage and voltages when the self-contained refrigerator or freezer is running and not running.
- Check and record voltages and amperage of the defrost heaters.
- Check and record anti-sweat heaters, voltage, and amperage.
- Check controls for proper operation against the manufacturer's service literature.
- Check all fan set screws and tighten if needed.

Self-Contained Commercial Refrigerators and Freezer Maintenance

When performing maintenance on the self-contained commercial equipment, always follow the manufacturer's recommendations for the proper chemicals to use on the equipment. To ensure a long life of service, proper sanitation, and minimum maintenance costs, these self-contained commercial refrigerators and freezers should be washed thoroughly on the inside and cleaned thoroughly on the outside and debris removed on a weekly basis.

SAFETY NOTE *Before performing maintenance on the appliance, turn off the electricity to the equipment, remove the food product from within, and place it into another refrigerator or freezer. When cleaning the inside cabinet, do not allow the cleaning chemicals or the cloth to come in contact with the food product or electrical components.*

Glass, glass door, and Plexiglas maintenance and cleaning should be performed daily and according to the manufacturer's recommendations provided in the service literature.

Exterior surfaces should be cleaned weekly with a mild detergent and warm water. Do not use abrasives to clean the exterior, it might damage the finish.

Remove any debris that might be clogging the waste water outlet and condensate drain pan (Figure 14-34). Debris sitting on the condensate pan heater might cause a burning smell. Clean the condensate drain pan, removing any debris and scale.

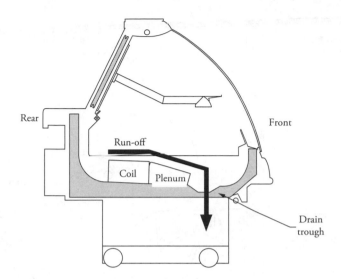

FIGURE 14-34
Arrow indicates the condensate flow towards the drain trough on its ways to the condensate pan.

🔱 Remove the honeycomb discharge grille (Figure 14-35*a* and 14-35*b*), use a spray detergent and a soft bristle brush to clean the honeycomb, and then rinse off with water and let it dry. Be careful, the honeycomb is brittle and it can break easily. After cleaning, re-install in reverse order of removal.

🔱 The evaporator coil is covered to keep food fluids from entering the coil area. Gain access to the evaporator coil, lift or remove the fan plenum (Figure 14-36), rinse with hot water. Be careful not to overflood the drain system. A wet/dry vacuum will come in handy for this procedure to remove excess water. Some cases are connected to the store drainage

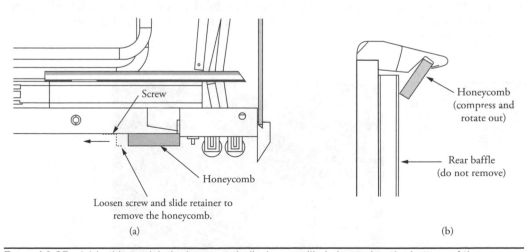

FIGURE 14-35 (*a*) In this model, the honeycomb discharge grille is located on the bottom of the case; (*b*) this model shows the honeycomb located above the rear baffle on the air discharge duct.

FIGURE 14-36
Gaining access to the
evaporator coil for
cleaning.

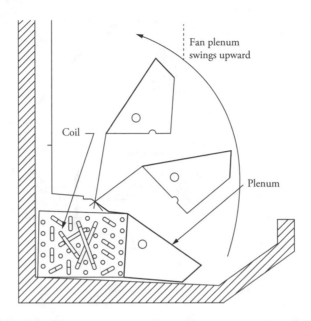

FIGURE 14-36
Gaining access to the
evaporator coil for
cleaning.

system, allowing for continuous water flow, without overfilling the condensate pan. Always
use cold water when cleaning evaporator coil area. Use a mild detergent and water solution
to clean the area around the evaporator coil area. When necessary, water and baking soda
solution can be used to remove odors from the case. After cleaning is complete, reinstall the
fan plenum, making sure the plenum is reinstalled correctly.

⬩ To clean the condenser coil (Figures 14-37 and 14-38), gain access per the
manufacturer's instructions in the service literature. Use a bush and a wet/dry vacuum to
clean the dust off the coil. Next, hook the vacuum hose up to the discharge side of the
vacuum, blow out the condenser area of any dust or debris. Then, reverse back the vacuum
hose to suck up dust and debris that was blown away from the condensing unit.
Reassemble in the reverse order of removal.

FIGURE 14-37
Dust and dirt covering
the condenser, fan
motor, and fan blade.

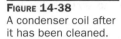

FIGURE 14-38
A condenser coil after it has been cleaned.

After the maintenance is performed, turn on the electricity and check the equipment for proper operation for 24 hours before placing the food into the self-contained refrigerator or freezer.

Suggested Cleaning and Maintenance Frequency Schedule

Table 14-1 will assist the owner on the suggested frequency of cleaning and maintenance of self-contained commercial refrigerators and freezers. Clean and sanitize using

Maintenance Frequency	Case Component
Annually	Case interior and evaporator coil
Annually	Honeycomb discharge grille
Daily	Check self-contained refrigerator or freezer for proper temperatures for the proper foods displayed.
Daily	Glass, glass door, and Plexiglas
Daily	Clean shelve(s) of food/fluid spillage
Daily	Check self-contained case for proper operation
Monthly	Return air grille
Quarterly	Condenser coil
Quarterly	Condenser unit area
Quarterly	Condensate pan and heater
Weekly	Exterior surfaces

TABLE 14-1   Frequency of Cleaning and Maintenance of Self-Contained Commercial Refrigerators and Freezers

chemicals recommended by the equipment manufacturer as stated in the service literature. Individual cleaning schedules must take into account local environment and usage, as well as all applicable health codes. Always read the manufacturer's service literature that came with the equipment before attempting any preventive maintenance procedures.

Commercial Ice Machines

This chapter will only cover the basics needed to perform the preventive maintenance on commercial ice machines. Each preventive maintenance procedure has been rated by a degree of complexity or simplicity. For the owner of the product, you will have to decide whether you can perform the preventive maintenance procedure or if you must call a professional to complete the preventive maintenance procedures on the product. Following are symbols that are used in this chapter to define who should perform the preventive maintenance on the commercial ice machine.

Expert	Call a certified HVAC service Company
Hard	Service technician
Moderate	Business Owner
Easy	Home Owner

The business owner will need hand tools and a flashlight to gain access and to see into the commercial ice machine.

The chapter will not cover diagnostics, any electrical repairs, or replacement procedures of any sealed-system components. This is a specialized area, and only an EPA-certified technician can repair the sealed system of self-contained refrigerators or freezers. The certified technician has the proper training, tools, and equipment required to make the necessary repairs to the sealed system. For the new HVAC/R student, the information in this chapter will be of great importance in your studies to become a certified air conditioning/refrigeration technician. An individual who is not certified to service the sealed system could raise the risk of personal injury, as well as property damage. The uncertified individual could void the manufacturer's warranty on the product if he or she attempts entry into the sealed system.

Commercial Ice Machines

Commercial ice is divided into three categories; ice cubes, flake ice, and nugget ice. The ice cube is pure ice. This means the ice cubes are free of impurities, chlorine, air, and odors are also eliminated. What remains in the ice cube is pure water frozen and is totally clear.

The flake ice is a soft ice, cloudy, and has impurities embedded within the ice. This type of ice is used for icing salads, fish, and poultry cases. The flake ice consists of 70% ice and 30% water. I'm sure you have seen this when you went to the supermarket. The nugget ice is similar to the flake ice, and it is larger and harder. The nugget ice consists of 80% ice and 20% water and it is more compressed than flake ice. It is used in ice dispensers, fish display, salad bars, and in fountain drinks and its appearance is cloudy.

Basic Cube Ice Machine Operation

With the ice machine turned on, with fresh water in the water trough (Figure 15-1), water is recirculating (Figure 15-2a and 15-2b), and during the freeze cycle, water is circulating over the evaporator plate(s) where ice cubes are formed (Figures 15-3 to 15-6). The compressor will stay on throughout the entire freeze and harvest cycles. The condenser fan motor will cycle on and off depending on which model is being serviced. When the ice has formed to a certain preset thickness, the harvest cycle begins (Figure 15-7). The water purge valve will open, allowing the water pump to purge the remaining water from the water trough, along with the impurities and sediment. The hot-gas solenoid valve is activated, allowing hot-gas refrigerant to go directly to the evaporator plate(s), heating the evaporator plate(s) and breaking the bond between the evaporator and the slab of ice (Figures 15-8 to 15-12). Once the ice is free from the evaporator plate(s), the ice slab is held together by a thin bridge between all the cubes (Figure 15-13). Next, the ice slab drops into the storage bin where it breaks into cubes (Figure 15-14). The harvest cycle has ended. The storage bin controls the amount of ice production on some models, with a bin thermostat that senses when the bin is full. On other models, with the storage bin full, the last slab of ice that begins to fall into the storage bin, stops and rests on the ice in the bin, holding open the water curtain, turning off the ice machine. As ice is removed from the storage bin, the last ice slab will fall into the bin, and ice production will begin again. Figure 15-15 identifies the component locations for an air-to-air type commercial ice machine. Figure 15-16 identifies the component locations for

FIGURE 15-1
Water entering the water trough in a commercial ice machine.

(a)   (b)

FIGURE 15-2 (a) Water pump recirculating the water after the water fill valve turns off; (b) the water pump is reticulating the cold water, on its way to the evaporator plate.

FIGURE 15-3 Water circulating water over the evaporator plate. You can see the ice formation within the cavities of the evaporator plate.

FIGURE 15-4 The ice formation is filling up the cavities in the evaporator plate.

FIGURE 15-5   The freeze cycle is almost complete.

FIGURE 15-6   The freeze cycle is about completed.

FIGURE 15-7   The water has contacted the ice thickness probe, the harvest cycle begins.

FIGURE 15-8   The ice has seperated from the evaporator plate.

FIGURE 15-9   The weight of the ice will allow the ice slab to fall into the bin.

FIGURE 15-10   The ice slab is about to leave the evaporator plate.

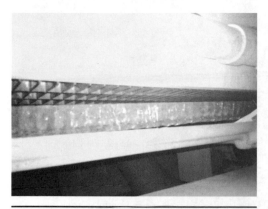

FIGURE 15-11 The ice slab is falling downward.

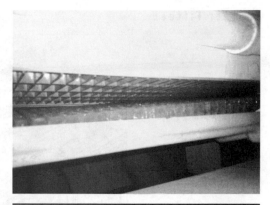

FIGURE 15-12 The ice slab is on it way to the storage bin.

FIGURE 15-13
The ice cubes are held together by the ice bridge that connects them together.

FIGURE 15-14
Ice cubes in the storage bin. The freeze cycle will begin to make ice again. Until the storage bin is full.

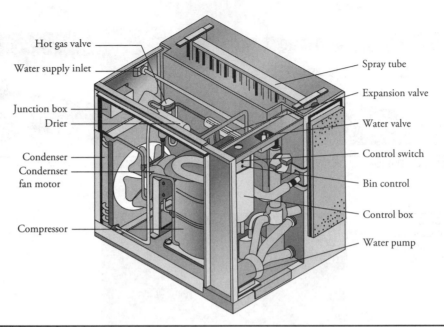

FIGURE 15-15   A typical air-to-air type commercial ice machine.

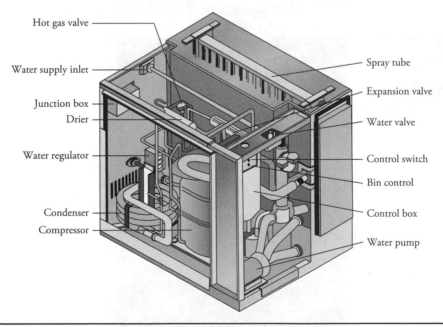

FIGURE 15-16   A typical air-to-water type commercial ice machine.

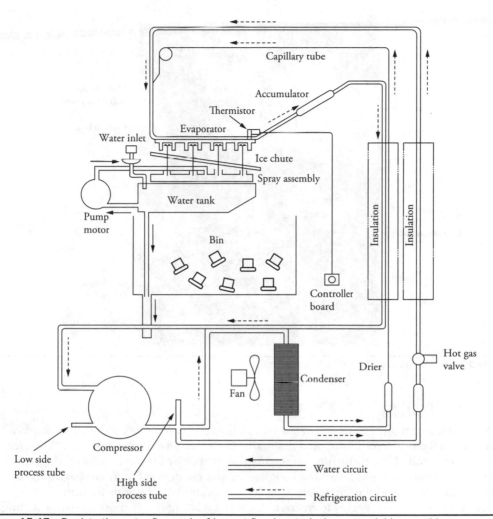

Figure 15-17 Depicts the water flow and refrigerant flow in a typical commercial ice machine.

an air-to-water-source type of commercial ice machine. Figure 15-17, shows the water flow and the refrigerant flow within the commercial ice machine. Figures 15-15 to 15-17 aid the service technician when diagnosing any problems within the ice machine.

Basic Flaker Ice Machine Operation

Water will enter a reservoir through the float valve and is gravity-fed into the evaporator barrel (Figure 15-18). The water entering the evaporator will fill up to the same level as the

Water inlet valve

Float located
inside the reservoir

Evaporator

Control box

Compressor

Gear reducer

V-Belt

Auger motor

Condenser
fan motor

Condenser coil

FIGURE 15-18   A typical commercial flaker ice machine.

water in the reservoir. The float located within the reservoir shuts off the water supply valve when the reservoir fills up and maintains the water level in the reservoir. When ice production is called for, by turning on the main switch or by the bin control closing, the flaker ice machine begins to run. The compressor and condenser fan motor begins to run after a slight delay and the temperature in the evaporator barrel begins to drop. As the water in the evaporator begins to freeze to the inner walls of the evaporator barrel, an auger motor turning a belt-driven gear reducer, turns the auger inside of the evaporator barrel, pushing the flaked ice upward, forcing the ice out of the top of the evaporator barrel through the delivery chute (Figures 15-19 and 15-20). The flaker ice machine does not incorporate a harvest cycle. The flaker ice machine will continuously run, producing flake ice. The flake ice is on its way to the storage bin for consumption. As the flake ice enters the storage bin, make-up water enters the bottom of the evaporator. This cycle continues until the ice storage bin is filled up and the ice machine turns off.

Basic Nugget Ice Machine Operation

The nugget ice machine (Figure 15-21) runs similar to the flake ice machine. When the flake ice leaves the evaporator barrel, it goes through an extruding head that tightly compress the ice, hitting the breaker, breaking off nugget size cubes. The nuggets enter the storage bin for consumption.

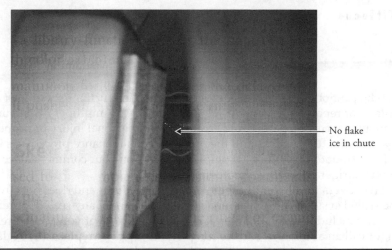

FIGURE 15-19 At the beginning of the flake ice machine freeze operation. Ice is beginning to freeze to the auger.

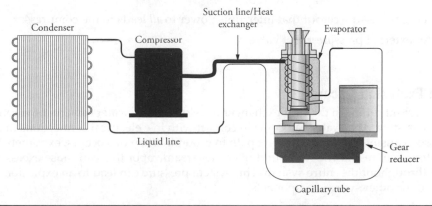

FIGURE 15-20 Flake ice will begin to travel from the auger to the storage bin.

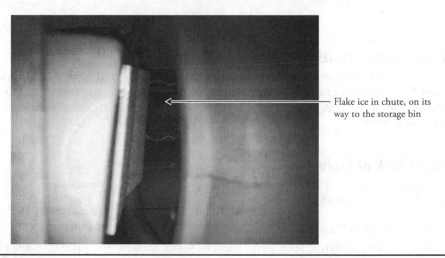

FIGURE 15-21 A typical nugget ice machine refrigeration circuit.

Safety First

Any person who cannot use basic tools or follow written instructions should *not* attempt to install, maintain, or repair any commercial ice machines. Any improper installation, preventive maintenance, or repairs could create a risk of personal injury or property damage. If you do not fully understand the preventive maintenance procedures in this chapter, or if you doubt your ability to complete the task on your commercial ice machine, please call your refrigeration service professional.

This chapter covers the preventive maintenance procedures and does not cover how to diagnose the sealed system. The actual repair or replacement of any sealed-system component is not included in this chapter. It is recommended that you acquire refrigerant certification (or call an authorized service company) to repair or replace any sealed-system component, as the refrigerant in the sealed system must be recovered properly.

Before continuing, take a moment to refresh your memory on the safety procedures in Chapter 1.

Prevention of Water-Utilizing System Explosions

In certain water-utilizing refrigeration systems, water can leak into the refrigerant side of the system. This can lead to an explosion of system components, including but not limited to the compressor. If such an explosion occurs, the resulting blast can kill or seriously injure anyone in the vicinity.

Systems at Risk of Explosion

Water-utilizing systems that have single-wall heat exchangers may present a risk of explosion. Such systems may include

- Water-source heat pump/air conditioning systems
- Water-cooling systems, such as ice makers, water coolers, and juice dispensers

Water-utilizing systems that have single-wall heat exchangers present a risk of explosion, unless they have one of the following:

- A high-pressure cutout that interrupts power to *all* leads to the compressor
- An external pressure-relief valve

How an Explosion Occurs

If the refrigerant tubing in the heat exchanger develops a leak, water can enter the refrigerant side of the system. This water can come in contact with live electrical connections in the compressor, causing a short circuit or a path to ground. When this occurs, extremely high temperatures can result. The heat buildup creates steam vapor that can cause excessive pressure throughout the entire system. This system pressure can lead to an explosion of the compressor or other system components.

Commercial Ice Machines in General

Much of the preventive maintenance information in this chapter covers commercial ice machines in general, rather than specific models, in order to present a broad overview of preventive maintenance techniques. The illustrations that are used in this chapter are for demonstration purposes only, to clarify the description of how to perform preventive maintenance on these appliances. They in no way reflect on a particular brand's reliability.

Treating the Water

In the freeze cycle, as the water passes over the evaporator freeze plate, the impurities in the water are rejected and only the pure water will stick to the plate. The more dissolved solids that are present in the water, the longer the freezing cycle. Bicarbonates, which are found in the water, are the most troublesome of all impurities. These impurities can cause

- Scaling on the evaporator freeze plate
- Clogging of the water distributor head
- The water valve and many other parts in the water system may clog up

If the impurities become too concentrated in the water system, they can cause cloudy cubes and/or mushy ice.

All of the water system parts that come in direct contact with the water might become corroded if the water supply is high in acidity. The water might have to be treated in order to overcome problems with the mineral content. The most economical way to treat the water supply is with a polyphosphate feeder. This feeder is installed in the water inlet supply to prevent scale buildup. This will require less frequent cleaning of the ice maker. To install one of these feeders, follow the manufacturer's recommendations in order to treat the water satisfactorily.

Preventive Maintenance Checklist

The following checklist will assist the certified HVAC/R service technician in performing a proper inspection on commercial ice machines. By properly maintaining the ice machine, the efficiency of the appliance will be operating at or above the 90% range of the rated ice production with a reduction in energy use. Always check the manufacturer's literature that comes with the commercial ice machine if there are additional preventive maintenance checks that are not listed on this checklist.

The following checklist is for preventive maintenance provided by a service technician on commercial ice machines:

Cube ice machines.

- Check exterior of ice machine for damage.
- Check air filter.
- Check ice machine for proper installation.

- Check ambient and storage bin temperatures.
- Clean ice machine and sanitize every 6 months or more frequently depending on water conditions or environment.
- Clean and sanitize the inside of the storage bin. Also, clean the outside of the storage bin.
- Clean the exterior of the ice machine.
- Check the ice bridge thickness.
- Check the water level in the water trough.
- Clean the condenser section for proper air flow.
- Check for water, refrigerant, or oil leaks.
- Check the water valve inlet screen, clean or replace as necessary.
- Check the bin thermostat for proper adjustment.
- For water cooled models, check the water valve for proper adjustment.
- Check all electrical connections for tightness.
- Check all wiring for burning, pinched, cut, broken, and loose wire nuts.
- Check all controls for proper operation.
- Check UVC lights.
- Check ice production.
- Check any other adjustments needed per the instructions in the service literature for the model you are servicing.
- Check voltage and amperage. Compare the readings against the data plate on the ice machine and manufacturer's service literature.
- Replace water filter(s) in the supply line if needed.

Flaker and nugget ice machines.

- Check exterior of ice machine for damage.
- Check air filter.
- Check ice machine for proper installation.
- Check ambient and storage bin temperatures.
- Clean ice machine and sanitize every 6 months or more frequently depending on water conditions or environment.
- Clean and sanitize the inside of the storage bin. Also, clean the outside of the storage bin.
- Clean the exterior of the ice machine.
- Check the float reservoir for mineral build-up.
- Check the auger drive motor amperage; this will determine if the ice machine needs cleaning.
- Inspect the bearing and auger for wear.

- Replace water filter(s) in the supply line if needed.
- Check the water valve inlet screen, clean or replace as necessary.
- Check the water level in the float tank.
- Clean the condenser section for proper air flow.
- Check for water, refrigerant, or oil leaks.
- Check the bin thermostat for proper adjustment.
- For water cooled models, check the water valve for proper adjustment.
- Check all electrical connections for tightness.
- Check all wiring for burning, pinched, cut, broken, and loose wire nuts.
- Check all controls for proper operation.
- Check UVC lights.
- Check ice production.
- Check any other adjustments needed per the instructions in the service literature for the model you are servicing.
- Check voltage and amperage. Compare the readings against the data plate on the ice machine and manufacturer's service literature.
- On water cooled ice machines, check the water-regulating valve for proper adjustment as described in the manufacturer's service literature.
- Check the thermostatic expansion valve (TXV) bulb to make sure it is properly secured and properly installed.
- Check the auger motor for lubrication if needed.
- Check the V-belt for wear and for proper tension according to the manufacturer's specifications listed in the service literature.

Cleaning and Sanitizing Commercial Ice Machines

Minerals separate from the water and accumulate on the evaporator plate or evaporator-barrel and water-system components. These minerals form what is known as *scale*. When the surfaces are dry, the scale becomes more visible. The color of the scale depends on the type of minerals in the water supply to the ice machine. When the scale is an off white in color, it is an indication of lime or calcium. When the scale is a rusty red color, it is an indication of iron. Algae or slime appearing in the ice machine indicates microbial growth; this has to be cleaned before someone gets sick from drinking or eating the ice. If scale is allowed to form on the evaporator plate or evaporator barrel, the ice will stick to the plate or barrel, reducing ice production. It can also cause poor ice quality. The ice remaining on the evaporator plate after a harvest will accumulate, causing a freeze up. The scale build up will also cause problems in the water-circulating system. The following cleaning and sanitizing procedures will alleviate this problem. Always check with the manufacturer's service literature for the model you are servicing for any additional instructions on cleaning or sanitizing the commercial ice machine.

Typical Cleaning Instructions for a Cubed Ice Machine

Cleaning of the cubed ice machines is usually performed every 6 months, unless the local water conditions require the service technician to clean the ice machine more often. The following maintenance procedure will assist the service technician in properly cleaning the ice machine.

1. Gain access to the ice machine by removing the panels.

2. When the ice machine has completed a harvest, turn off the ice machine.

3. Remove all of the ice from the storage bin.

4. Use only an approved cleaner recommended by the ice machine manufacturer for cleaning. Add the recommended amount of ice machine cleaner mixed with water, to the water trough, according to the instructions listed in the service literature and listed by the ice machine cleaner manufacturer's instructions on the bottle. If the ice machine cleaner is not mixed correctly, heavy foaming will occur and the possibility of damaging the ice machine components.

5. Initiate the wash cycle, by placing the "ICE/OFF/WASH" switch to the "WASH" position. To remove the mineral deposits, allow the ice machine cleaner to circulate for 15 to 20 minutes.

6. Terminate the wash cycle by sliding the "ICE/OFF/WASH" switch to the "OFF" position. Inspect the evaporator plate and water spillway, by removing the curtain, check and see if residue has been removed.

7. Activate the purge function on the ice machine, this will allow the ice machine cleaner to be flushed down the drain and diluted with fresh clean incoming water.

8. Wipe down evaporator plate, spillway, and any other water transport surfaces with a clean soft cloth to remove any remaining residue. Clean all water distribution tubing with a bottle brush to remove any residue remaining. Inspect the distributor for restricted orifice holes, clean as needed. If there is still residue remaining within the ice machine, repeat steps 4 to 8.

9. Clean any other components as listed in the service literature for the model you are servicing according to the manufacturer's instructions. Make sure all surfaces are clean from scale and slime before proceeding to the sanitizing stage.

10. Preparing to sanitize the ice machine involves adding sanitizing solution to fill the water trough, to the top of the trough. For the model ice machine you are servicing, only add the recommended amount of sanitizing solution, to the water trough, according to the instructions listed in the service literature and listed by the sanitizing solution manufacturer's instructions on the bottle. Next, move the "ICE/OFF/WASH" switch to the "WASH" position and let the solution circulate for 10 to 15 minutes. This will be a good time to inspect the ice machine for any leaks. While the ice machine is sanitizing the water-distribution system and the evaporator plate wipe down all other ice machine splash areas, including the interior of the storage bin, and curtain with sanitizing solution. Make sure that all parts are reinstalled including any fasteners and thermostat bulbs (if the model you are servicing has one).

11. Engage the purge solenoid valve and keep it energized until all of the sanitizing solution has been flushed down the drain. Next, turn on the ice machine water supply and continue to purge the sanitizing solution for another 2 to 3 minutes.

12. You are now ready to place the ice machine back into ice production. Make sure the air filter is clean. Also, replace the inlet water supply filter.

13. Place the **"ICE/OFF/WASH"** switch to the **"ICE"** position and replace all panels on the ice machine.

14. You're not done yet, wait until the ice machine produces 2 to 3 ice harvests, then dispose of the ice.

15. Check total ice machine operation. Now you are done.

Typical Cleaning Instructions for a Flaker Ice Machine

Cleaning the flake ice machines is usually performed every 6 months, unless the local water conditions require the service technician to clean the ice machine more often. The following maintenance procedure will assist the service technician in properly cleaning the ice machine:

1. Gain access to the ice machine by removing the panels.

2. Remove all of the ice from the storage bin.

3. Turn the ice machine and the water supply to the float off.

4. Use only an approved cleaner recommended by the ice machine manufacturer for cleaning. Use the recommended amount of ice machine cleaner according to the instructions listed in the service literature and listed by the ice machine cleaner manufacturer's instructions on the bottle. If the ice machine cleaner is not mixed correctly, heavy foaming will occur and the possibility of damaging the ice machine components.

5. Turn on the ice machine, remove the float reservoir cover, and add ice machine cleaning solution to the reservoir.

6. As the ice machine makes ice, keep adding the ice machine cleaning solution until the amount (as specified by the ice machine manufacturer service instructions) is used up.

7. Turn the ice machine off.

8. Preparing to sanitize the ice machine involves adding the specified amount of sanitizing solution to the reservoir while the ice machine is producing ice. For the model ice machine you are servicing only add the recommended amount of sanitizing solution, to the reservoir, according to the instructions listed in the service literature and listed by the sanitizing solution manufacturer's instructions on the bottle.

9. Turn the ice machine off.

10. Replace the float reservoir cover back on the reservoir and turn on the water supply to the float.

11. Next, turn on the ice machine and let it produce ice for the next 15 to 20 minutes.

12. Turn off the ice machine and discard the ice from the storage bin.

13. Clean the inside of the storage bin with warm soapy water including the bin door, and door frame. Then, rinse out the storage bin and rinse the remaining parts.

14. Using sanitizing solution, wipe down all areas of the storage bin, door, and door frame and rinse.

15. Finally, turn on the ice machine again. You have completed the cleaning and sanitizing of the ice machine.

Self-Contained Ice Machine Performance Data

There are a number of factors that can affect the overall performance of the ice machine.

Installation

- Installation in a dusty environment will cause the ice machine to operate continuously and have a low ice production rate.
- Installation in a greasy atmosphere will cause the ice machine to operate in high temperature conditions with low to no ice production.
- Installation next to a high-temperature appliance will cause the ice machine to operate in high-temperature conditions with low to no ice production.

Water Temperature

- Higher water temperatures will increase the operating time of the ice machine.
- Decreased ice production in a 24-hour period.

Ambient Temperature

- High temperatures will decrease ice production.
- Ice storage bin melting ice above 25%.

Maintenance

- Restricted air movement through the condenser coil and front grille will cause low to no ice production.

Other

- If the ice cubes are thicker than specified in the manufacturer's service literature the ice production will decrease and the operating time will increase.

Troubleshooting Water and Its Effect on Ice Making

When making ice cubes in a self-contained ice machine, the quality of the ice should be defined as solid, clear, and free of taste or odor. The ice machine can provide this if the quality of the water is pure. Tables 15-1 and 15-2 will show some of the ingredients that can affect ice cube production:

Ingredient	Effect	Solution
Algae	Horrible taste and odor	1. Install a carbon filter.
Minerals: Calcium Magnesium Potassium Sodium	Cloudy ice cubes	1. Check the water flow for a restriction (water valve or water line kinked). 2. Change the water source. 3. Install a polyphosphate feeder or water softener.

TABLE 15-1   Ingredients That Affect Ice Cube Quality

Ingredient	Effect	Solution
Chlorine Iron Manganese	Staining of ice machine components	1. Clean ice machine with ice machine cleaner and ice machine sanitizer. 2. Install a water softener and iron filter.
Permanent water hardness Calcium or magnesium Chlorides Nitrates Sulfates	Scale will form on ice machine components	1. Abrasive cleaning. Cleaning the ice machine more frequently. 2. Polyphosphate feeder or water softener reduces or eliminates the need for abrasive cleaning.
Temporary water hardness Calcium or magnesium Carbonates	Scale will form on ice machine components	1. Clean ice machine with ice machine cleaner and ice machine sanitizer. 2. Polyphosphate feeder or water softener reduces the frequency of cleaning the ice machine by 50%.

TABLE 15-2   Ingredients That Affect the Ice Machine Operation and Quantity of Ice Production

Ice Usage Guide

Table 15-3 indicates the amount of ice needed with a 20% safety factor figured into the amount for customers in restaurants, cocktail bars, hotels and motels, hospitals, and cafeterias. This guide can also be used to properly size an ice machine for a customer.

Application Type	Typical Daily Usage of Ice	Ice Needed for 100 Customers	Ice Needed for 250 Customers	Ice Needed for 500 Customers	Ice Needed for 1000 Customers	Ice Needed for 1500 Customers
Restaurant	1.5 lb ice per meal sold	180 lb	450 lb	900 lb	1800 lb	2700 lb
Cocktail bar	3 lb ice per seat	360 lb	900 lb	1800 lb	3600 lb	5400 lb
Glass of water	6 oz ice per 12 oz glass	45 lb	113 lb	225 lb	450 lb	675 lb
Salad bar	35 lb ice per cubic foot	35 lb ice per cubic foot	35 lb ice per cubic foot	35 lb ice per cubic foot	35 lb ice per cubic foot	35 lb ice per cubic foot
Beverage only	5-oz ice per 7–10 oz cup	38 lb	94 lb	188 lb	375 lb	563 lb
Beverage only	8 oz ice per 12–16-oz cup	60 lb	150 lb	300 lb	600 lb	900 lb
Beverage only	12-oz ice per 18–24 oz cup	90 lb	225 lb	450 lb	900 lb	1350 lb
Guest ice	5 lb per hotel/motel room	600 lb	1500 lb	3000 lb	6000 lb	9000 lb
Hotel catering	1 lb per person	120 lb	300 lb	600 lb	1200 lb	1800 lb
Patient ice	10 lb per person	1200 lb	3000 lb	6000 lb	12000 lb	18000 lb
Cafeteria	1 lb per person	120 lb	300 lb	600 lb	1200 lb	1800 lb

TABLE 15-3 Ice Usage Guide

16
CHAPTER

Troubleshooting

This chapter will assist the service technician in diagnosing problems with the HVAC or refrigeration equipment. For the home owner and business owner, please call a professional to come out and do the repairs on your equipment.

Each troubleshooting procedure has been rated by a degree of complexity or simplicity. Following are symbols that are used in this chapter to define who should perform the diagnostics and repairs on the HVAC and refrigeration.

Expert

Hard

Moderate

Easy

Call a certified HVAC service Company

Service technician

Business Owner

Home Owner

Repairing the sealed system in a HVAC or refrigeration system is a specialized area, and only an EPA-certified technician can repair the sealed system on air conditioning or refrigeration equipment. The certified technician has the proper training, tools, and equipment required to make the necessary repairs to the sealed system. For the new HVAC student, the information in this chapter will be of great importance in your studies to become a certified air conditioning or refrigeration technician. An individual who is not certified to service the sealed system could raise the risk of personal injury, as well as property damage. The uncertified individual could void the manufacturer's warranty on the product if he or she attempts entry into the sealed system.

Safety First

Any person who cannot use basic tools or follow written instructions should *not* attempt to install, maintain, or repair any air conditioning or refrigeration equipment. Any improper installation, preventive maintenance, or repairs could create a risk of personal injury or property damage. If you do not fully understand the troubleshooting procedures in this chapter, or if you doubt your ability to complete the task on the HVAC or refrigeration equipment, please call your HVAC service professional.

This chapter covers the troubleshooting procedures only, the actual repair or replacement of any sealed-system component is not included in this chapter. It is recommended that you acquire refrigerant certification (or call an authorized service company) to repair or replace any sealed-system component, as the refrigerant in the sealed system must be recovered properly.

Before continuing, take a moment to refresh your memory on the safety procedures in Chapter 1.

HVAC and Refrigeration Systems in General

Much of the troubleshooting procedures information in this chapter covers commercial air conditioners and refrigeration products in general, rather than specific models, in order to present a broad overview of troubleshooting techniques. The illustrations that are used in this chapter are for demonstration purposes only, to clarify the description of how to troubleshoot the HVAC or refrigeration equipment. They in no way reflect on a particular brand's reliability.

Basic Troubleshooting Techniques

There are many different ways to diagnose a problem, but all of them basically use the same reasoning of deduction:

- Where does the consumer think the malfunction is located within the HVAC or Refrigeration product?
- Where is the actual problem located within HVAC or Refrigeration product?
- Are there any related problems with HVAC or Refrigeration product?
- How can the problem with HVAC or Refrigeration product?

For example, the consumer states that the HVAC unit does not heat and believes that the electric heating element is bad. The actual problem might be a voltage problem, a bad fuse or circuit breaker, bad fuse link, no control voltage, bad heating elements, faulty thermostat, or improper thermostat settings.

When checking the heat, you might notice that the thermostat settings are set for cooling instead of heating. Thus, the actual problem was that the thermostat settings were not positioned correctly. The related problem is "How did the thermostat setting move to the cooling position?" This leads to the question: "Does the consumer know how to operate the thermostat?" To solve this problem, you will have to instruct the consumer in the proper operation of the thermostat.

All HVAC equipment goes through a certain sequence of events. Understanding the proper operation and this sequence as indicated in the use and care manual or the service manual is beneficial when diagnosing the HVAC and refrigeration equipment.

Product Failure

Given the information about the product's problem, information and diagnostic figures and charts from this book, and the information you have read in the use and care manual, service manual, and installation instructions as the servicer, you will be able to perform the following steps in sequence to diagnose and correct a malfunction. The basic steps to follow when diagnosing HVAC and refrigeration equipment:

1. *Verify the complaint.* Ask the consumer what symptoms were caused by the problem with the HVAC or refrigeration equipment. When was the last time a service company performed a preventive maintenance check up on the equipment?

2. *Check for external factors.* For example, is the HVAC or refrigeration equipment installed properly, does the product have the correct voltage, etc.?

3. *Check for physical damage.* Look for internal and external physical damage. Any damage will prevent the HVAC or refrigeration equipment from functioning properly. Two examples are a broken fan blade paddle in the condenser unit or a damaged cabinet that prevents the doors from closing properly.

4. *Check the controls.* The controls must be set to the proper settings. If the controls are not set correctly, the HVAC or refrigeration equipment might not function properly or complete its cycle.

5. *Operate the product.* Operate the HVAC or refrigeration equipment, and let it run through its cycle. Check the cycle operation against the operational sequence of events that is listed in the installation instructions or service manual.

6. *Is the product operating properly?* If it is, explain to the consumer how to operate the HVAC or refrigeration equipment according to the manufacturer's specifications.

7. *The product is not operating properly.* If the HVAC or refrigeration equipment is not operating properly, proceed to locate which component has failed. Check the diagnostic charts that are listed in this book or the service manual to assist you in the correct direction to take.

Diagnosis and Correction Procedure

When diagnosing a problem with HVAC or refrigeration equipment, use your five senses to determine the condition of the product. This will help in analyzing and defining the problem:

- *Example #1.* When turning on the HVAC, there is a smell of something burning. You can track down the location of the burning smell and, therefore, discover which part has failed.

- *Example #2.* When turning on a refrigerator, unusual noises are heard coming from underneath the product. Turn off the refrigerator, and attempt to track down where the noises are coming from.

Along with your hand tools, there are a variety of test meters that can assist you in analyzing and defining the problem. This is the sequence of events to follow when servicing HVAC or refrigeration equipment:

- *Unplug the product.*   Change the range setting on the multimeter to voltage. Check the supply voltage. If there is an uncertainty, check the name plate rating for the correct voltage rating; this is located on the product. When diagnosing a component failure, there are three types of circuit failure—the open circuit, the grounded circuit, and the short circuit—all of which are thoroughly explained in Chapter 6.

- *Gain access.* Only remove the panels and screws necessary to gain access to where the suspected component failure is located.

- *Isolate and/or remove the defective part.* Using the multimeter, isolate and/or remove the part, set the range to ohms, and check for component failure. This is thoroughly explained in Chapter 6.

- *Install the new component.* When you find a defective part, replace it with a new original part. Reconnect all the wires in their original places.

- *Reattach all panels and screws.* Close the HVAC or refrigeration product, and reattach all panels and screws.

- *Test the product for proper operation.* Turn on the supply voltage to the HVAC or refrigeration equipment and test it. The product operation and performance must be according to the manufacturer's specifications as listed in the service literature.

Technician's, Diagnostic Guide

Before a technician begins to service HVAC or refrigeration equipment, he or she must check the following conditions before the service call begins, during the diagnostic phase of the service call, and after the service call is completed:

- Make sure that there is the correct supply of voltage at the HVAC or refrigeration equipment to operate correctly.

- Check and see if a fuse has blown or a circuit breaker has tripped. If a blown fuse or circuit breaker has tripped, do not replace the fuse or reset the circuit breaker until you have tested the compressor, for an open winding, short, or a ground.

- Check and see if the HVAC or refrigeration equipment has been installed according to the manufacturer's instructions.

- All meter readings should be made with a multimeter (VOM or DVM) with a sensitivity of 20,000 Ω/V DC or greater.

- Locate the service literature for the product model you are servicing. You can call the product manufacturer or look up the information online. The installation instructions that come with the product contain the wiring diagram (also provided within the cabinet or posted on a panel) and other useful information needed to diagnose the fault and complete the repairs.

- On electronic models you will need the actual product service manual for the model you are working on to properly diagnose the product. The service manual will assist you in understanding the fault codes, voltages, and adjustments. The service manual will also assist you in the step-by-step process of testing the electrical circuits correctly.

- During the diagnostic part of the service call, check all connections first before you replace any parts. Check for loose connections or burnt wires. During the testing phase you will have to disconnect and reconnect wires. Be careful not to pull on the wires; you might pull the wire off the terminal connector.

- Check all wire harness connectors first. Inspect each wire harness connector, and make sure that there are no loose or broken wires on the connectors. Make sure that all wires are pressed into the connector properly.

- Resistance checks must be performed with the electricity turned off to the HVAC or refrigeration equipment.
- Voltage checks must be performed with the electricity turned on and tested for the correct voltage present.
- All refrigerant levels or sealed-system repairs must be checked by an EPA-certified service technician.
- When you complete the service call, make sure that the HVAC or refrigeration equipment operates according to the manufacturer's specifications.

Diagnostic Electrical Flowcharts

A flowchart is a diagnostic tool that helps the technician to diagnose HVAC or refrigeration equipment problems. It provides you with a step-by-step process to solve the problem. Choose the flowchart needed, start at the beginning, follow the arrows to the next box, and so on, until you solve the problem. The following are for diagnosing the electrical circuits of a product:

- Direct current voltage diagnostic flow chart (Figure 16-1)
- 24-V AC diagnostic flow chart (Figure 16-2)

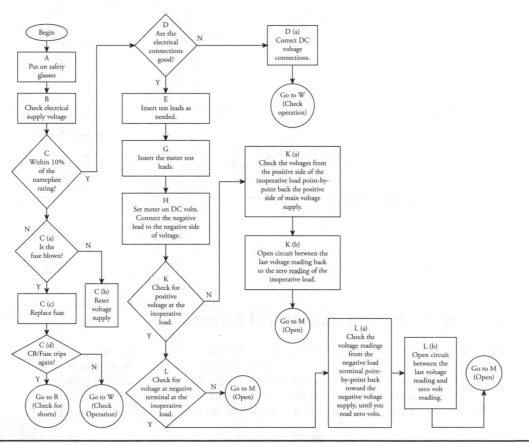

Figure 16-1   Direct current diagnostic flowchart. (*continued*)

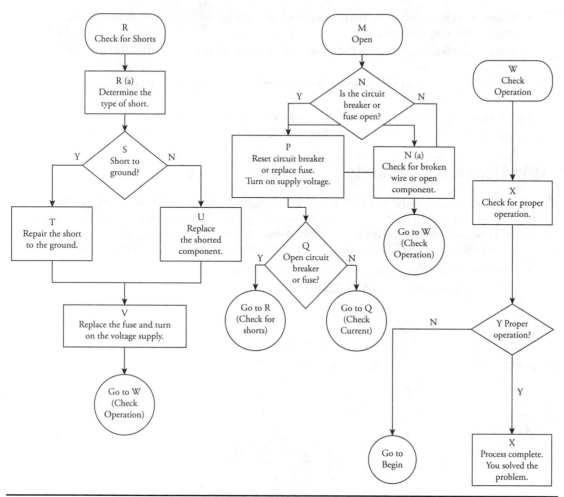

Figure 16-1 Direct current diagnostic flowchart.

- 120-V AC diagnostic flow chart (Figure 16-3)
- 240-V AC diagnostic flow chart (Figure 16-4)

Troubleshooting the Refrigeration Sealed System

When servicing the air conditioning or refrigeration system, do not overlook the simple things that might be causing the problem. Before you begin testing and servicing an HVAC or refrigeration system, check for the most common faults. Inspect all wiring connections for broken or loose wires. Check and see if all components are running at the proper times during the cycle. Test the temperatures in the HVAC or refrigeration system inlet and outlet grilles.

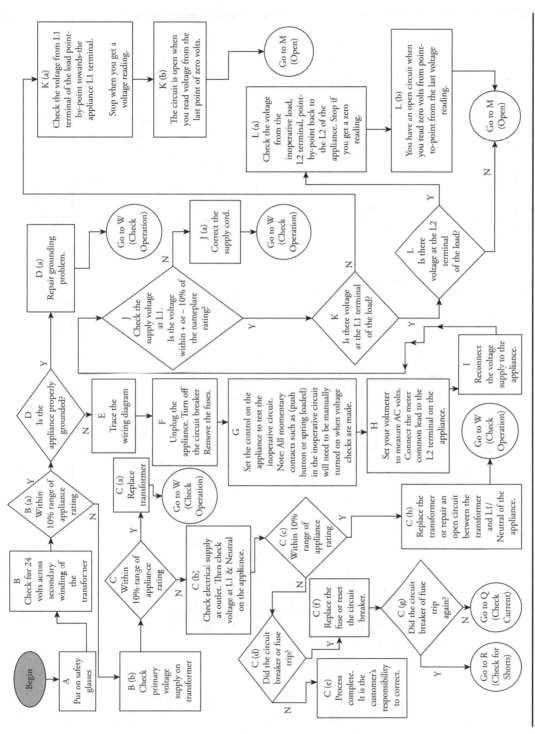

FIGURE 16-2 24-V AC diagnostic flowchart. (*continued*)

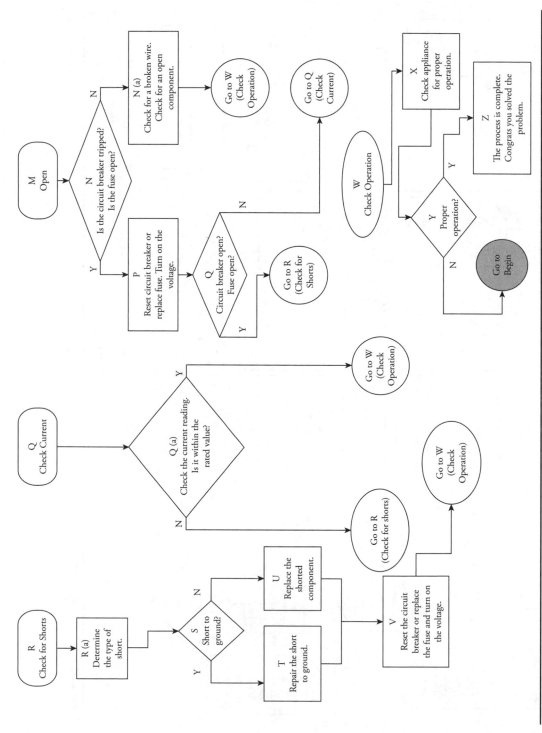

Figure 16-2 24-V AC diagnostic flowchart.

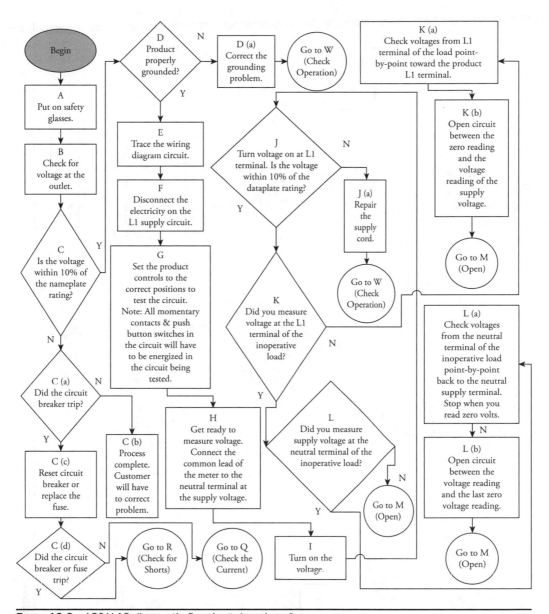

FIGURE 16-3   120-V AC diagnostic flowchart. (*continued*)

Tables 16-1 and 16-2 are intended to serve as aids to assist the technician in determining if there is a sealed-system malfunction, providing that all other components are functioning correctly within the HVAC or refrigeration system that mimic a sealed-system failure. You must rule out everything else before you enter the sealed system.

NOTE *Before you enter a sealed system, you must be an EPA certified to do so.*

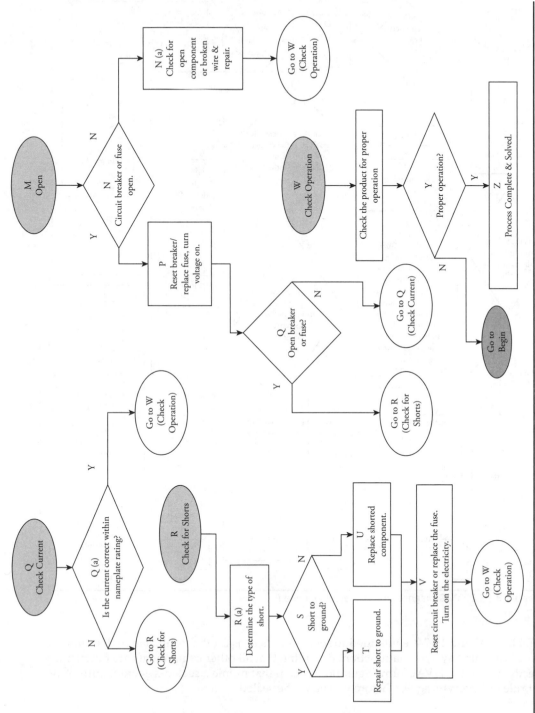

Figure 16-3 120-V AC diagnostic flowchart.

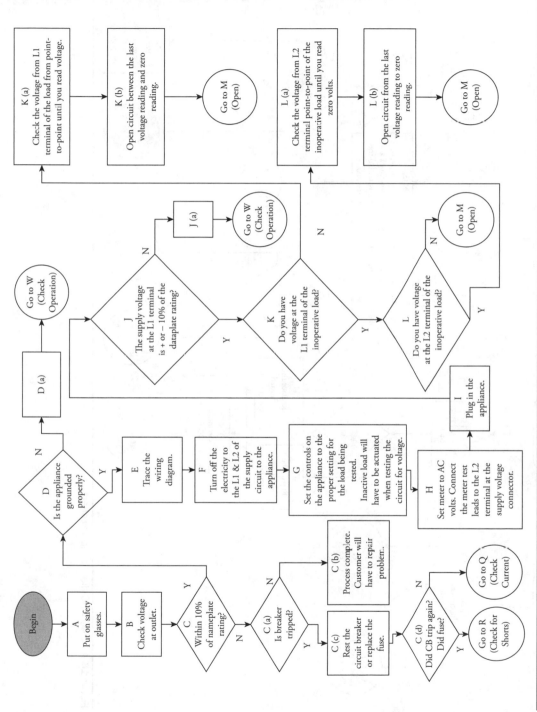

Figure 16-4 240-V AC diagnostic flowchart. *(continued)*

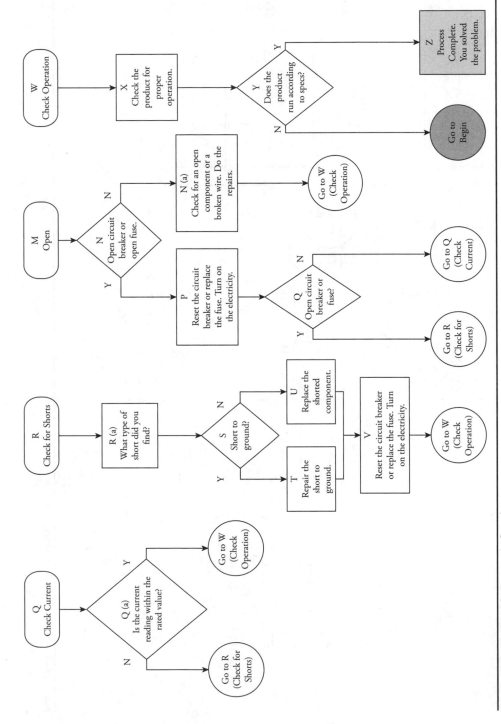

Figure 16-4 240-V AC diagnostic flowchart.

Refrigeration Sealed System Diagnosis Chart

Sealed-System Condition	Suction Line Pressure (TECHNICIANS ONLY)	Liquid Line Pressure Line Pressure (TECHNICIANS ONLY)	Suction Line to Compressor	Compressor Discharge Line	Condenser Coil	Capillary Tube	Evaporator Coil	Frost Line	Amperage or Wattage	Pressure Equalization Rate
Normal Operation	Normal pressure readings	Normal pressure readings	Slightly below room temperature	**WARNING** Very hot to the touch	**WARNING** Very hot to the touch	Warm	Cold coil and maintaining proper temperatures	Suction line from inside box not frozen	Normal meter readings	Normal
Sealed-System Overcharged	Higher than normal pressure readings	Higher than normal pressure readings	Heavily frosted; may be very cold to the touch	Between slightly warm and hot	Between hot and warm to the touch	Cool; below room temperature	Cold coil and possibly not maintaing temperatures	All the way back to the suction line	Higher than normal meter readings	Normal to slightly longer
Sealed-System Undercharged	Pressure readings are lower than normal	Pressure readings are lower than normal reading	Warm to the touch; possibly near room temperature	**WARNING** Hot to the touch	The entire coil feels warm	Warm	Inlet to coil feels extremely cold while the outlet from the coil will be below room temperature	Partial	Lower than normal meter readings	Normal
Partial Restriction within Sealed System	Lower than normal pressure readings, possibly in a vacuum	Intermittent lower than normal reading	Warm to the touch; possibly near room temperature	**WARNING** Very hot to the touch	The top passes in the coil are warm and the lower passes are cool, near to room temperature	Feels like room temperature between cool and colder	Inlet to coil feels extremely cold while the outlet from the coil will be below room temperature	Intermittent / frost line will begin to grow in length	Lower than normal meter readings	Intermittent
Complete Restriction within Sealed System	Pressure readings are in a deep vacuum	Ambient readings	Feels the same as room temperature	Feels the same as room temperature	Feels the same as room temperature	Feels the same as room temperature	No refrigeration or air conditioning	None	Lower than normal meter readings	No equalization
Out of Refrigerant Possible Leak in System	The pressure reading will be from 0 PSIG to 30" vacuum	Atmospheric reacing	Feels the same as room temperature	Can feel like cool to hot	Feels the same as room temperature	Feels the same as room temperature	No refrigeration or air conditioning	Non-existent	Lower than normal meter readings	Normal
Low Capacity Compressor	Higher than normal pressure readings	Lower than normal readings	Cool to room temperature	Cooler than normal	Low	Warm to box temperature	Partial or half of the evaporator frost pattern	Partial to non-existent	Lower than normal	Quicker than normal

TABLE 16-1 Refrigeration Sealed-System Diagnosis Chart, with Capillary Tube for a Metering Device

Conditions That Mimic Sealed System Failures

Conditions	Amperage or Wattage	Condenser Coil Temperature	Frost Line	Compressor Discharge Line Temperature	Low Side Pressure (for Service Technicians Only)	High Side Pressure (for Service Technicians Only)	Fresh Food Compartment Temperature	Freezer Compartment Temperature
Plugged Condenser Coil	Higher than normal	Higher than normal	Full	Higher than normal	Higher than normal	Higher than normal	Warmer than normal readings	Warmer than normal readings
Blocked Condenser Fan Assembly	Higher than normal	Higher than normal	Full	Higher than normal	Higher than normal	Higher than normal	Warmer than normal readings	Warmer than normal readings
Blocked Evaporator Fan Assembly	Lower than normal	Lower than normal	Frost back to compressor	Lower than normal	Lower than normal	Lower than normal	Warmer than normal readings	Warmer than normal readings
Evaporator Coil Iced Up (Defrost Failure)	Lower than normal	Lower than normal	Frost back to compressor	Lower than normal	Lower than normal	Lower than normal	Warmer than normal readings	Warmer than normal readings
High Head Load	Higher than normal	Higher than normal	Full	Higher than normal	Higher than normal	Higher than normal	Warmer than normal readings	Warmer than normal readings
High Ambients	Higher than normal	Higher than normal	Full	Higher than normal	Higher than normal	Higher than normal	Warmer than normal readings	Warmer than normal readings
Damper Failed Closed	Lower than normal	Lower than normal	Full	Lower than normal	Lower than normal	Lower than normal	Warmer than normal readings	Cooler than normal readings
Damper Failed Open	Slightly higher than normal	Slightly higher	Full	Normal	Slightly higher than normal	Normal	Cooler than normal	Normal to slightly warmer readings

TABLE 16-2   Conditions That Mimic a Sealed-System Failure

Possible Problem	Discharge Pressure	Suction Pressure	Superheat Reading	Subcooling Reading	Compressor Amperage
Sealed system under charged	Low	Low	High	Low	Low
Sealed system over charged	High	Low	High	Low	Low
Incorrect air flow—evaporator	Low	Low	Low	High	Low
Dirty condenser Coil or poor airflow	High	High	High	High	High
Outside ambient temperature—low	Low	Low	Low	High	Low
Sealed system restriction—liquid line drier	Low	Low	High	High	Low
Metering device (TXV) —lost charge	Low	Low	High	High	Low
Metering device (TXV) —bulb sensing ambient temperature— poor insulation on bulb	High	High	Low	Low	High
Inefficient compressor	Low	High	High	High	Low

TABLE 16-3   HVAC/Refrigeration Diagnostic Chart with a TXV Installed

HVAC/Refrigeration Sealed System Troubleshooting Chart

Table 16-3 indicates the refrigerant pressures, superheat, subcooling, and compressor amperage for a refrigeration sealed system that has a thermostatic expansion valve installed in the refrigeration circuit.

Troubleshooting Sealed-System Problems in a Room Air Conditioner

If you suspect a sealed-system malfunction, be sure to check out all external factors first. These include

- Thermostats.
- Compressor.
- Overload on the compressor.
- Fan motor.
- Evaporator and condenser coils getting good air circulation.
- Air conditioner installation.

- Make sure the heater(s) is not on at the same time as cooling.
- Make sure that the reversing valve and solenoid coil are operating properly (heat pump models).

After eliminating all of these external factors, you will then systematically check the sealed system. This is accomplished by comparing the conditions found in a normally operating air conditioner. These conditions are

- Room temperature
- Wattage
- Condenser temperature
- Evaporator inlet sound (gurgle, hiss, etc.)
- Evaporator cooling pattern
- High-side pressure[2]
- Low-side pressure[2]
- Pressure equalization time

One thing to keep in mind: No single indicator is conclusive proof that a particular sealed-system problem exists. Rather, it is a combination of findings that must be used to definitively pinpoint the exact problem.

Low-Capacity Compressor

 Symptoms of a low-capacity compressor in the sealed system are

- Temperatures in the room or area will be above normal.
- The wattage and amperage will be below normal, as indicated on the model/serial plate.
- The temperature of the condenser coil will be below normal.
- At the evaporator coil, you will hear a slightly reduced gurgling noise.
- The evaporator coil will show a normal cooling pattern.
- The high-side pressure will be below normal, and the low-side pressure will be above normal.[2]
- The pressure equalization time might be normal or shorter than normal.

Refrigerant Leak

 Symptoms of a refrigerant leak in the sealed system are

- Temperatures in the room or area will be below normal.
- The wattage and amperage will be below normal, as indicated on the model/serial plate.
- The condenser coil will be cool to the touch at the last pass, or even as far as midway through the coil.

- At the evaporator coil, you will hear a gurgling noise, a hissing noise, or possibly an intermittent hissing or gurgling noise.
- The evaporator coil will show a frost pattern in the lower rungs of the coil.
- The high- and low-side pressures will be below normal.[2]
- The pressure equalization time might be normal or shorter than normal.

Overcharged Air Conditioner

 If the sealed system is overcharged, the symptoms are

- The room temperature will be higher than normal.
- The wattage and amperage will be above normal, as indicated on the model/serial plate.
- The temperature of the condenser coil will be above normal.
- At the evaporator coil, you will hear a constant gurgling noise—generally, a higher sound level than normal.
- The evaporator coil will show a full frost pattern. If you remove the cover, you will possibly see the suction line frosted back to the compressor.
- The high- and low-side pressures will be above normal.[2]
- The pressure equalization time will be normal.

Slight Restriction

 Symptoms of a slight restriction in the sealed system are

- The room temperature will be below normal.
- The wattage and amperage will be below normal, as indicated on the model/serial plate.
- The temperature of the condenser coil will be slightly below normal.
- At the evaporator coil, you will hear a constant gurgling noise and a low sound level.
- The evaporator coil cooling pattern will be receded.
- The high- and low-side pressures will be below normal.[2]
- The pressure equalization time will be longer than normal.

Partial Restriction

 Symptoms of a partial restriction in the sealed system are

- The room temperature will be higher than normal.
- The wattage and amperage will be below normal, as indicated on the model/serial plate.
- The temperature of the condenser coil will be below normal more than halfway on the coil.

- At the evaporator coil, you will hear a constant gurgling noise and a considerably lower sound level.
- The evaporator coil cooling pattern will be considerably receded.
- The high- and low-side pressures will be below normal.[2]
- The pressure equalization time will be longer than normal.

Complete Restriction

 Symptoms of a complete restriction in the sealed system are

- The room temperature will be warm.
- The wattage and amperage will be considerably below normal, as indicated on the model/serial plate.
- The temperature of the condenser coil will be cool or at room temperature.
- At the evaporator coil, you will hear no sounds.
- The evaporator coil will not be cool.
- The high-side pressure will be equal to the pressure of refrigerant at room temperature.
- The low-side pressure will be in a deep vacuum.[2]
- There will be no pressure equalization time.

Moisture Restriction

 Symptoms of a moisture restriction in the sealed system are

- The room temperature will be above normal.
- The wattage and amperage will be considerably below normal, as indicated on the model/serial plate.
- The temperature of the condenser coil will be below normal.
- At the evaporator coil, you will hear a constant gurgle, low sound level, or no sound at all.
- The evaporator coil might have some frost on the evaporator inlet.
- The high-side pressure will be below normal.[2]
- The low-side pressure will be below normal or in a deep vacuum.[2]
- The pressure equalization time will be longer than normal or there will be no equalization at all.

Reversing Valve (Heat Pump Models)

In a straight-cool air conditioner, the refrigerant flows from the compressor discharge through the condenser coil, capillary tube, evaporator coil, and back through the suction line to the compressor. The ability of an air conditioner to reverse the direction of the rcfrigerant flow is achieved with a reversing valve installed in the refrigerant circuit. The reversing valve is designed as a single-port, double-throw valve. It turns the function of the evaporator coil from a heat absorption coil into a heat dissipation coil (condenser coil) (Figure 16-5). When the solenoid coil is energized, the valve will reverse the refrigerant flow. For cooling, in a heat-pump air conditioner, the refrigerant flows from the compressor

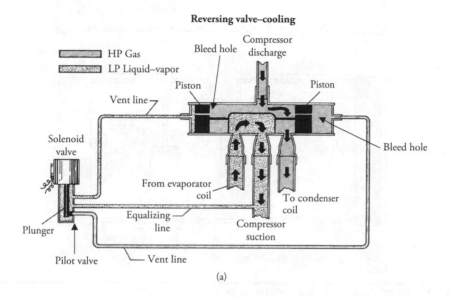

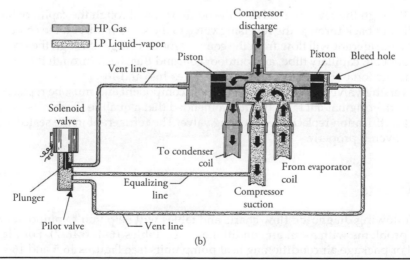

FIGURE 16-5   The refrigerant flow in a reverse-cycle valve. (*a*) The reversing valve solenoid coil is de-energized in the cooling mode; (*b*) the reversing valve solenoid coil is energized in the heating mode.

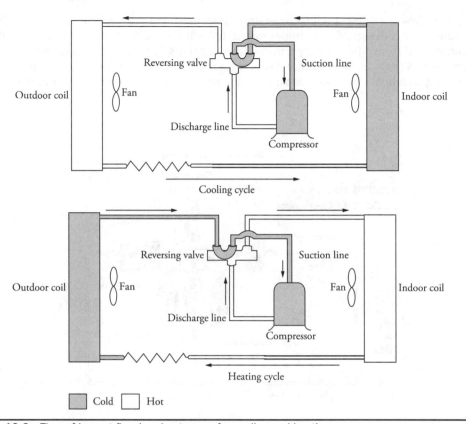

Figure 16-6 The refrigerant flow in a heat pump for cooling and heating.

discharge through the reversing valve to the outdoor coil, through the capillary tube and indoor coil, and back through the reversing valve to the suction port on the compressor. For heating, the refrigerant will flow from the compressor discharge; through the reversing valve, indoor coil, capillary tube, and outdoor coil; and then back through the reversing valve to the suction port on the compressor (Figures 16-6 and 16-7).

The reversing valve itself is a non-serviceable component and must be replaced with a duplicate of the original if it fails. It is recommended that a qualified technician with refrigerant certification replace the reversing valve. The refrigerant in the sealed system must be recovered properly.

Diagnostic Charts

The following diagnostic flow charts and tables will help you to pinpoint the likely causes of problems with room air conditioners (see Tables 16-4 to 16-7). For diagnostics on central or package air conditioning heat pump units (see Figures 16-5 and 16-7, and Tables 16-4 and 16-5).

FIGURE 16-7
(*a*) Illustrates the reversing valve in the heating mode; (*b*) illustrates the reversing valve in the cooling mode. The reversing valve is also known as a four-way valve.

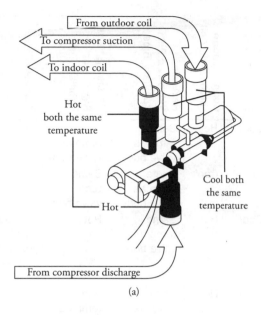

From outdoor coil

To compressor suction

To indoor coil

Hot both the same temperature

Hot

Cool both the same temperature

From compressor discharge

(a)

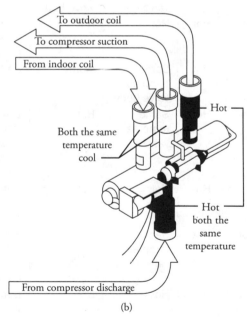

To outdoor coil

To compressor suction

From indoor coil

Both the same temperature cool

Hot

Hot both the same temperature

From compressor discharge

(b)

Reversing Valve Operating Condition	Discharge Line from the Compressor	Suction Line to the Compressor	Copper Line to the Inside Coil	Copper Line to the Outside Coil	Reversing Valve Left Pilot Capillary Tube	Reversing Valve Right Pilot Capillary Tube	Probable Cause	Solution
Normal cooling	Hot	Cool	Cool	Hot	Same temp as valve body	Same temp as valve body	Normal	None
Normal heating	Hot	Cool	Hot	Cool	Same temp as valve body	Same temp as valve body	Normal	None
Reversing valve will not shift from cooling to heating	Hot	Cool	Cool	Hot	Same temp as valve body	Hot	Bleeder hole blockage or leaking piston	Replace reversing valve
	Hot	Cool	Cool	Hot	Same temp as valve body	Same temp as valve body	Vent tube blockage	Replace reversing valve
	Hot	Cool	Cool	Hot	Hot	Hot	Both pilot valve ports are open	Replace reversing valve
	Warm	Cool	Cool	Hot	Same temp as valve body	Warm	Bad compressor	Replace compressor
Reversing valve will begin to shift and it will not complete the reversal	Hot	Warm	Warm	Hot	Same temp as valve body	Hot	Low on refrigerant or valve body damage	Check refrigerant charge or replace
	Hot	Warm	Warm	Hot	Hot	Hot	Both pilot valve ports are open	Replace reversing valve
	Hot	Hot	Hot	Hot	Same temp as valve body	Hot	Stuck valve or valve body damage	Replace reversing valve
	Hot	Hot	Hot	Hot	Hot	Hot	Both pilot valve ports are open	Replace reversing valve

TABLE 16-4   Reversing Valve Touch Test Chart

Reversing Valve Operating Condition	Discharge Line from the Compressor	Suction Line to the Compressor	Copper Line to the Inside Coil	Copper Line to the Outside Coil	Reversing Valve Left Pilot Capillary Tube	Reversing Valve Right Pilot Capillary Tube	Probable Cause	Solution
Fluctuating heating	Hot	Cool	Hot	Cool	Same temp as valve body	Same temp as valve body	Internal valve leaking at piston	Replace reversing valve
	Hot	Cool	Hot	Cool	Valve body warmer than usual	Valve body warmer than usual	Internal valve leaking at piston	Replace reversing valve
Reversing valve will not shift from heating to cooling	Hot	Cool	Hot	Cool	Same temp as valve body	Same temp as valve body	High refrigerant pressure differential	Restart air conditioner and recheck
	Hot	Cool	Hot	Cool	Hot	Same temp as valve body	Bleeder hole blockage	Replace reversing valve
	Hot	Cool	Hot	Cool	Hot	Same temp as valve body	Internal valve leaking at piston	Replace reversing valve
	Hot	Cool	Hot	Cool	Hot	Hot	Defective pilot valve	Replace reversing valve
	Warm	Cool	Warm	Cool	Warm	Same temp as valve body	Bad compressor	Replace compressor

Note: Before replacing the reversing valve, turn the air conditioner off for a few minutes, and restart the cycle. On some occasions, the reversing valve may return to normal operation.

TABLE 16-5 Reversing Valve Touch Test Chart

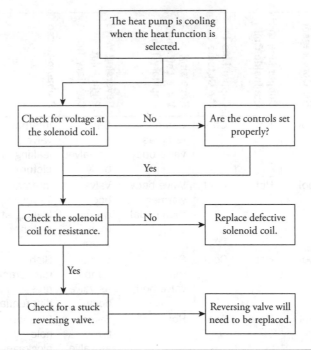

TABLE 16-6   The Diagnostic Flowchart: The Air Conditioner Is Cooling When the Heat Function Is Selected

---

Water Source Heat Pump Thermostatic Expansion Valve (TXV) Troubleshooting Chart

The diagnostic chart in Table 16-8 will help you pinpoint the likely causes of problems with water source heat pump air conditioning units:

Air Source Heat Pump Air Conditioning Unit Troubleshooting Chart

The diagnostic chart in Table 16-9 will help you pinpoint the likely causes of problems with air-source heat pump air conditioning units:

Static Pressure Reading

Figures 16-8 to 16-10 demonstrate the locations on HVAC equipment where the technician should test total static pressures. Measuring static pressure is a fundamental part of diagnosing how the airflow and duct system are performing. If the static pressure is too

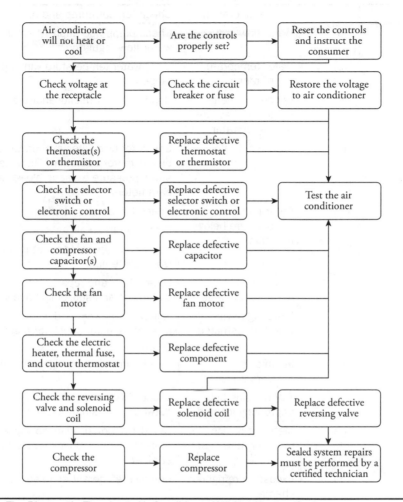

TABLE 16-7 The Diagnostic Flowchart: Air Conditioner Will Not Heat or Cool

high, you have evidence of problems with the HVAC system causing low airflow. Examples, blockage in ducts, closed dampers, improper duct transitions, offsets or kinked flex duct, evaporator coil issues, or filter(s) issues. Low static pressure can also mean there is a problem with the HVAC system and ductwork system. Low pressure may indicate leaking ductwork or plenums, missing air filters, low fan speed or separated ductwork.

Motor Troubleshooting Charts

The following diagnostic chart (Tables 16-10 and 16-11) will help you pinpoint the likely causes of problems with motors in HVAC and refrigeration equipment.

Symptom	Heating Mode	Cooling Mode	Cause	Solution
Low refrigerant suction pressure	Yes	No	Normal operation	Check temperatures and pressures
	Yes	No	Water flow reduced	Check pump operation, strainer or filter, water flow regulator
	Yes	No	Water temperature out of range	Bring water temperature within proper range
	Yes	No	Condenser heat exchanger has water scaling in water	Conduct water quality analysis
	No	Yes	Reduced air flow	Check for dirty filter, evaporator coil, blower motor operation. The external static pressure exceeds blower operating parameters.
	Yes	Yes	Return air temperature below minimum	Space temperature too cold.
				Excessive fresh air.
	No	Yes		Supply air bypassing to return air stream (zone systems).
	Yes	Yes	Insufficient refrigerant charge	Locate leak and repair. Conduct superheat and subcooling analysis.
	Yes	Yes	Metering device failed or restricted	Failed TXV power head, capillary tubing, or sensing bulb. Check for plugged TXV strainer.
High refrigerant superheat	Yes	Yes	Insufficient refrigerant charge	Locate and repair refrigerant leak.
	Yes	Yes	TXV sensing bulb is improperly located	Locate TXV sensing bulb on suction line between reversing valve and compressor.
	Yes	Yes	Metering device failed or restricted	Failed TXV power head, capillary tubing, or sensing bulb. Check for plugged TXV strainer.
High refrigerant subcooling	Yes	Yes	Excessive refrigerant charge	Remove refrigerant as needed.
			Metering device failed or restricted	Failed TXV power head, capillary tubing, or sensing bulb. Check for plugged TXV.
Evaporator coil, tubing, or TXV frosting	Yes	No	Normal operation	This may occur when the entering water temperature is at or below minimum temperature.
	Yes	Yes	Insufficient refrigerant charge	Locate and repair refrigerant leak.
	Yes	Yes	Metering device failed or restricted	Failed TXV power head, capillary tubing, or sensing bulb. Check for plugged TXV.
Equalizer line frosting or condensing	Yes	Yes	Metering device failed or restricted	Failed TXV power head, capillary tubing, or sensing bulb. Check for plugged TXV.

TABLE 16-8   Water Source Heat Pump Thermostatic Expansion Valve (TXV) Troubleshooting Chart

Symptom	Heating Mode	Cooling Mode	Cause	Solution
Low refrigerant suction pressure	Yes	No	Normal operation	Check temperatures and pressures
	Yes	Yes	Evaporator fan motor not running or incorrect speed setting.	Check evaporator fan motor and capacitor. ECM motors, check speed settings, motor, and the motor module.
	Yes	Yes	Condenser fan motor not running.	Check fan motor, fan blade, run capacitor. ECM motors, check speed settings, motor, and the motor module.
	No	Yes	Reduced air flow	Check for dirty filter, evaporator coil, blower motor operation. The external static pressure exceeds blower operating parameters.
	Yes	Yes	Return air temperature below minimum	Space temperature too cold.
				Excessive fresh air.
	No	Yes	Insufficient refrigerant charge	Supply air bypassing to return air stream (zone systems).
	Yes	Yes		Locate leak and repair. Conduct superheat and subcooling analysis.
	Yes	Yes	Metering device failed or restricted	Failed TXV power head, capillary tubing, or sensing bulb. Check for plugged TXV strainer. Check piston for blockage.
	Yes	Yes	Insufficient refrigerant charge	Locate and repair refrigerant leak.
High refrigerant superheat	Yes	Yes	TXV sensing bulb is improperly located	Locate TXV sensing bulb on suction line between reversing valve and compressor.
	Yes	Yes	Metering device failed or restricted	Failed TXV power head, capillary tubing, or sensing bulb. Check for plugged TXV strainer.
	Yes	Yes	Excessive refrigerant charge	Remove refrigerant as needed.
High refrigerant subcooling			Metering device failed or restricted	Failed TXV power head, capillary tubing, or sensing bulb. Check for plugged TXV.
	Yes	No	Normal operation	This may occur when the entering water temperature is at or below minimum temperature.
Evaporator coil, tubing, or TXV frosting	Yes	Yes	Insufficient refrigerant charge	Locate and repair refrigerant leak.
	Yes	Yes	Metering device failed or restricted	Failed TXV power head, capillary tubing, or sensing bulb. Check for plugged TXV.
	Yes	Yes	Metering device failed or restricted	Failed TXV power head, capillary tubing, or sensing bulb. Check for plugged TXV.
Equalizer line frosting or condensing	Yes	Yes	Failed or restricted metering device	Failed TXV power head, capillary tubing, or sensing bulb. Check for plugged TXV.

TABLE 16-9   Air Source Heat Pump Air Conditioning Unit Troubleshooting Chart

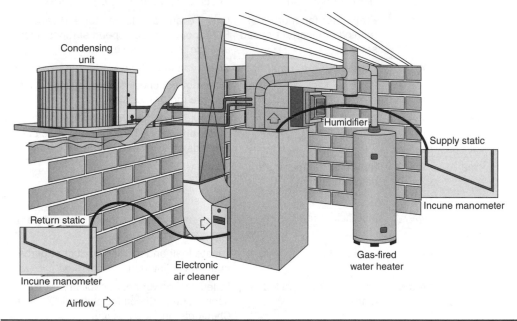

FIGURE 16-8   Upflow total static-pressure reading locations.

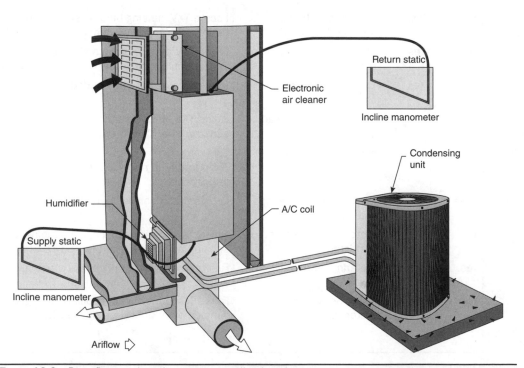

FIGURE 16-9   Downflow total static-pressure reading locations.

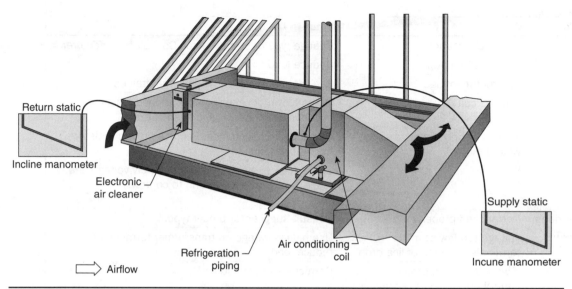

Return static

Incline manometer

Electronic
air cleaner

Supply static

Incune manometer

Refrigeration
piping

Air conditioning
coil

⟹ Airflow

FIGURE 16-10   Horizontal right total static-pressure reading locations.

Problem	Cause	What to Do
Motor fails to start	Blown fuses or tripped circuit breaker	Replace fuses at least 125% of nameplate amperes. Reset circuit breaker.
	Overload trips	Check and reset overload in starter.
	Improper supply voltage	Check supply voltage supplied to motor agrees with the nameplate and load factor
	Improper line connections	Check connections with wiring diagram supplied with the motor
	Open circuit in winding or starter switch	Check for loose wiring connections; also check and see if starter switch is closed inside of motor.
	Mechanical failure	Check to see if motor and drive turn freely. Check bearings and lubrication.
	Short circulated stator	Indicated by blown fuses or tripped circuit breaker. The motor must be rewound or replaced.
	Poor stator coil connection	Remove the end bells, locate with multimeter.
	Rotor defective	Look for broken bars or end rings. Replace motor.
	Motor may be overloaded	Reduce load.
	If three-phase motor, one phase may be open	Check lines for open phase with a multimeter.
	Defective capacitor	Check for short circuit, grounded or open capacitor. Replace if necessary.
	Worn or sticking brushes on repulsion motors	Check for wear and correct brush pressure. Clean commutator if dirty.

TABLE 16-10   General Motor Troubleshooting Chart (*continued*)

Problem	Cause	What to Do
Motor stalls	Wrong application	Change type or size of motor. Consult manufacturer.
	Overloaded motor	Reduce load
	Low motor voltage	See that nameplate voltage is maintained.
	Open circuit	Fuses blown or circuit breaker tripped, check overload, relay, stator, and push-buttons.
	Incorrect control resistance of wound motor	Check control sequence. Replace broken resistors. Repair open circuit.
Motor turns and then dies	Power failure	Check for loose connections to supply voltage line, fuses, or circuit breaker and to control.
Motor does not come up to speed	Not applied properly	Consult supplier for proper type.
	Voltage too low at motor terminals because of line drop	Use higher voltage on transformer terminals or reduce load
	If wound rotor, improper control operation of secondary resistance	Correct secondary control
	Starting load too high	Check load motor is supposed to carry and start.
	Broker rotor bars	Lock for cracks near the rings. A new rotor may be required. Repairs are temporary.
Motor takes too long to accelerate	Excess loading	Reduce load.
	Poor circuit	Check for high resistance.
	Defective squirrel cage rotor	Replace with new rotor.
	Applied voltage is too low	Get electric company to increase power tap.
Wrong rotation	Wrong sequences of phases	Reverse connections at motor or switchboard.
Motor overheats while running under load	Overload	Reduce load.
	Dirt build up in motor	Clean and remove dirt from motor.
	Motor may have one phase open	Check that all leads are well connected.
	Grounded coil	Locate and repair.
	Unbalanced terminal voltage	Check for faulty leads, connections, and transformers.
	Shorted stator coil	Repair or replace and check wattage.
	Faulty connection	Indicated by high resistance.
	High voltage	Check terminals with multimeter for high voltage.
	Low voltage	Check terminals with multimeter for high voltage.
	Rotor rubs stator bore	If not poor machining replace bearings.

TABLE 16-10   General Motor Troubleshooting Chart (*continued*)

Problem	Cause	What to Do
Motor vibrates after corrections have been made.	Motor misaligned	Realign.
	Weak foundations	Strengthen base.
	Coupling out of balance	Balance coupling.
	Driven equipment unbalanced	Rebalance driven equipment.
	Defective ball bearing	Replace ball bearing.
	Bearings not in line	Line up properly.
	Balancing weights shifted	Rebalance rotor.
	Wound rotor coils	Rebalance rotor.
	Poly phase motor running single phase. Excessive end play	Check for open circuit, adjust bearing or add washer.
Unbalanced line current on poly phase motors during normal operation	Unequal voltage at terminals	Check leads and connections.
	Single phase operation	Check for open contacts.
	Poor rotor contacts in control wound rotor resistance	Check control devices.
	Brushes not in proper position	See that brushes are properly seated and shunts in good condition.
Scraping noise	Fan rubbing air shield	Remove interference.
	Fan striking insulation	Clear fan.
	Loose on bedplate	Tighten holding bolts.
Noisy operation	Air gap not uniform	Check and correct bracket fits or bearing.
	Rotor unbalance	Rebalance.
Hot bearings general	Bent or sprung shaft	Straighten or replace shaft.
	Excess belt pull	Decrease belt tension.
	Pulleys too far away	Move pulley closer to the motor bearing.
	Pulley diameter too small	Use larger pulleys.
	Misalignment	Correct by realignment of drive.
Hot bearings sleeve	Oil grooving obstructed by dirt	Remove bracket or pedestal with bearing and clean bearing housing and oil grooves; renew oil.
	Oil too heavy	Use recommended lighter oil.
	Oil too light	Use recommended heavier oil.
	Too much end thrust	Reduce thrust induced by drive, or supply external means to carry thrust.
	Badly worn bearing	Replace bearing.
Hot bearings ball	Insufficient grease	Maintain proper amount of grease in bearing.
	Deterioration of grease or lubricant contaminated	Remove oil grease, wash bearings thoroughly in kerosene and replace with new grease.
	Excess lubricant	Reduce quantity of grease. Bearing should not be more than half filled.
	Overloaded bearing	Check alignment side and end thrust.
	Broken ball or rough races	Replace bearing, first clean housing thoroughly.

TABLE 16-10   General Motor Troubleshooting Chart

Symptoms	Fault	Possible Causes	Corrective Action	Notes
Motor rocks slightly when starting	This is normal start-up for ECM/ICM motor	—	—	—
Motor won't start	No movement	• Manual disconnect switch is off or door switch open • Blown fuse or circuit breaker • Miswired 24 volt wires • Unseated pins in wiring harness connections • Bad motor/control module • Moisture present in motor or control module	• Check 115 voltage at motor • Check for 24 vac voltage at motor • Check low voltage connections • Check for unseated pins in connectors • Test temporary with a jumper between R-G	• Turn electricity off prior to repair. • Wait 5 minutes after turning off electricity to the motor, before servicing. This will allow the capacitors sufficient time to discharge • Handle ECM/ICM module with care.
	Motor rocks but won't start	• Loose motor mount • Blower wheel not tight on motor shaft • Bad motor/control module	• Check for loose motor mount • Make sure blower wheel is tight on motor shaft • Perform motor/control replacement check, ICM-2 motors only.	• Turn electricity off prior to repair. • Wait 5 minutes after turning off electricity to the motor, before servicing. This will allow the capacitors sufficient time to discharge • Handle ECM/ICM module with care.
Motor oscillates up and down while being tested off of blower	It is normal for motor to oscillate with no load on shaft.	—	—	—

Symptom	Possible Cause	Check	Safety
Motor starts, but runs erratically	• Variation in 115 Vac to motor • Unseated pins in wiring harness connections • Erratic CFM command from terminal • Improper thermostat connection or setting • Moisture present in motor/control module	• Check line voltage for variation • Check low voltage connections at motor, unseated pins in motor harness connectors • Check out system controls – Thermostat • Perform moisture check*	Turn off electricity prior to repair
"Hunts" or "Puffs" at high CFM (speed)	• Incorrect or dirty filter • Incorrect supply or return ductwork • Incorrect blower speed setting	• Does removing panel or filter reduce "puffing"? • Check/replace filter • Check/correct duct restrictions • Adjust to correct blower speed setting	Turn off electricity prior to repair
Motor stays at low CFM despite call for cool or heat CFM.	• 24 Vac wires miswired or loose • Check wiring at motor for loose connection • Fan in delay mode	• Check low voltage (Thermostat) wires and connections • Verify fan motor is not in delay mode – wait until delay is complete • Perform motor/control replacement check, For ICM-2 motors only.	• Turn electricity off prior to repair. • Wait 5 minutes after turning off electricity to the motor, before servicing. This will allow the capacitors sufficient time to discharge • Handle ECM/ICM module with care.
Motor stays at high CFM	• 24 Vac wires miswired or loose • Check wiring at motor for loose connection • Fan in delay mode	• Is fan motor in delay mode? Wait until delay time is complete. • Perform motor/control replacement check, For ICM-2 motors only.	• Turn electricity off prior to repair. • Wait 5 minutes after turning off electricity to the motor, before servicing. This will allow the capacitors sufficient time to discharge Handle ECM/ICM module with care.
Blower won't shut off	• Current leakage from controls into other terminals	• Check for Triac switched thermostat or solid state relay	Turn electricity off reior to repair.

TABLE **16-11** Troubleshooting Chart: ECM/ICM Variable Speed Blower Motors (*Continued*)

357

Symptoms	Fault	Possible Causes	Corrective Action	Notes
Excessive noise	Air noise	• High static creating high blower speed • Incorrect supply or return ductwork • Incorrect or dirty filters • Incorrect blower speed setting	• Check/replace filter • Check/correct duct restrictions • Adjust to correct blower speed setting	Turn off electricity prior to repair
	Noisy blower or cabinet	• Loose blower housing panels, etc. • High static creating high blower speed • Air leaks in ductwork, cabinetsm, or panels	• Check for loose blower housing, panels, etc. • Check for air whistling thru seams in ducts, cabinets, and panels • Check for cabinet/duct deformation	Turn off electricity prior to repair
	"Hunts" or "Puffs" at high CFM (speed)	• High static creating high blower speed • Incorrect or dirty filter • Incorrect supply or return ductwork • Incorrect blower speed setting	• Does removing panel or filter reduce "puffing"? • Check/replace filter • Check/correct duct restrictions • Adjust to correct blower speed setting	Turn off electricity prior to repair
Evidence of moisture	Motor failure or malfunction has occurred and moisture is present	• Moisture in motor/control module	• Replace motor and perform moisture check*	• Turn electricity off prior to repair. • Wait 5 minutes after turning off electricity to the motor, before servicing. This will allow the capacitors sufficient time to discharge • Handle ECM/ICM module with care.

*Moisture Check
• Connections are oriented "down" or as recommended by the equipment manufacturer.
• Arrange wiring harnesses with a "drip loop" under the motor.
• Check and make sure the condensate drain is not plugged up.
• Check for low airflow (too much latent capacity)
• Check for uncharged condition.
• Check and plug leaks in return ducts and cabinet.

TABLE 16-11 Troubleshooting Chart: for ECM/ICM Variable Speed Blower Motors

Faults That Can Be Felt	Effect on System Operation
Solenoid valve	
• Colder than the tubing ahead of the solenoid valve. (Solenoid valve sticks, partially open)	• Vapor in liquid line
• Same temperature as tubing ahead of solenoid valve. (Solenoid valve closed)	• System stopped via low-pressure control
Filter drier	
• Filter drier colder than tubing ahead of the filter drier. (Filter partially blocked with dirt on the inlet side of the filter drier)	• Vapor in the liquid line.
Faults That Can be Heard	**Effect on System Operation**
Regulators in suction line	
• Whining sound from evaporating pressure regulator [Regulator too large (sizing error)]	• Unstable operation
Compressor	
(a) Knocking sound on starting. (oil boiling).	(a) Liquid hammer. Risk of compressor damage.
(b) Knocking sound during operation (oil boiling) (Wear on moving parts).	(b) Liquid hammer. Risk of compressor damage.
Cool room	
• Defective alarm system (Lack of maintenance)	• Can give rise to personal injury
Faults That Can be Smelled	**Effect on System Operation**
Cold room	
• Bad smell in meat cold room.	• Leads to poor food quality and/or wastage.
(Air and humidity too high because evaporator too large or load too low)	

TABLE 16-12   Troubleshooting Chart: Commercial Refrigeration Faults That Can Be Felt, Heard, or Smelled

Commercial Refrigeration Faults That Can Be Felt, Heard, or Smelled

When diagnosing a problem with refrigeration equipment, use your senses to determine the condition of the product (Table 16-12). This will help in analyzing and defining the problem.

Endnotes

1. If you did not read resistance, it does not mean that the electric heater element is defective. After removing the heater assembly, test the thermal fuse and the heater limit control (thermostat mounted to heater frame) for continuity separately. The thermal fuse is a non-resetable fuse—it will open the heating circuit if the heating temperature

or amperage exceeds its rating, indicating that there may be other problems with the air conditioner (e.g., a defective fan motor, dirty filter, or dirty evaporator coil). The thermal fuse acts as the last line of defense if the heater limit control fails. The heater limit control will also open the circuit, limiting the heating, if there is a reduction in airflow. On some models, this device will reset, automatically, allowing the electric heater to come back on again. Both of these devices are installed in the heating circuit to prevent the heating element from causing a fire in the air conditioner and home or office. These safety devices must remain in the circuitry. Do not bypass these safety devices. Replace the defective component with a duplicate of the original.

2. If you open up the sealed system, you will void your warranty. The sealed system must be repaired by an authorized service company.

Where to Get Help

Keep careful records. Always put complaints in writing, and keep copies of all correspondence and service receipts. Be sure to ask for service receipts, even for no-charge, in-warranty calls.

NOTE DETAILS *when the problem was first noticed, when it was reported, and the servicing history (who serviced the air conditioning or refrigeration equipment, when, what was done, and how often service was required).*

If there are complaints about the air conditioning or refrigeration equipment, there are three steps to follow:

1. Read the use and care manual that comes with the product. Also check the circuit breaker, as well as fuses, pilots, and controls.

2. Call the service company authorized to fix the brand equipment. They have the training and equipment to deal with air conditioning or refrigeration service problems.

3. If not satisfied, contact the manufacturer's main customer relations office. The address and phone number are located in the use and care manual.

Product Recalls and the Internet

There are times when a product might have an electrical, gas, or mechanical issue that might be hazardous to the consumer. The U.S. Consumer Product Safety Commission (CPSC) was created to protect the consumer from serious injuries, unreasonable risks, death, and property damage.

The CPSC's web site (*http://www.cpsc.gov*) includes information on air conditioning or refrigeration equipment, and other product recalls. It also includes other information beneficial to the consumer. In addition, the following list provides a listing of companies, Web sites, and phone numbers for air conditioning, refrigeration, and other (companies involved in HVAC) manufacturers. You can also look in the use and care manual that comes with the product for the manufacturer's Web site. Remember, safety and education must be considered at all times when operating, maintaining, or repairing any air conditioning or refrigeration equipment.

Manufacturers

Company	Products	Web Address	Phone
Aaon H/C Products	Rooftop units	*www.aaon.com*	918.583.2266
Acme Fan	Fans/ventilation	*www.acmefan.com*	918.682.7791
Acutherm Therma-fuser	VAVs	*www.acutherm.com*	510.785.0510
A/C Store HVAC	Equipment/services	*www.acstore.com*	409.779.6700
Addison HVAC	Air handlers/heat pumps	*www.addison-hvac.com*	407.292.4400
Adobe Air, Inc.	Evaporative coolers	*www.adobeair.com*	602.257.0060
Advanced Radiant	Radiant heating	*www.advancedradiant.com*	206.783.4315
AERCO	Heating systems	*www.aerco.com*	800.526.0288
AESYS Technologies	Air handlers/boilers	*www.aesystech.com*	717.755.1081
Air Technology Inc.	Dehumidification	*www.airtechnologyinc.com*	800.822.8040
Air Turbine Propeller	Propeller fan blades	*www.airturbine.com*	724.452.9540
A-J Manufacturing Co.	Air distribution	*www.ajmfg.com*	800.247.5746
Ajax (Ace) Boiler Inc.	Boilers	*www.aboilerinc.com*	714.437.9050
Alabama Heat Exchanger	Heat exchangers	*www.heatexchanger.com*	334.653.1166
Alfa Lavel	Heat exchangers	*www.hvac.alfalaval.com*	804.222.5300
ALNOR Fume	Hood control	*www.alnor.com*	800.424.7427
Amana Htg. & A/C	AC units	*www.amana-hac.com*	877.254.4729
Amena (Thailand)	AC units	*www.amena-air.com*	662.517.1819
American Boiler	Boiler	*www.americanboiler.com*	800.388.1741
American Standard	(Trane) AC units	*www.amstd-comfort.com*	888.273.6397
AMTROL Inc.	Expansion tanks/air elimators	*www.amtrol.com*	401.884.6300
Anemostat	Air distribution	*www.anemostat.com*	310.835.7500
Annex Air	(Canada) energy recovery	*www.annexair.com*	819.475.3302
Allied Air	(Lennox) Armstrong air, air ease	*www.alliedair.com*	972.497.6670
Armstrong Air	(Lennox) Armstrong air, air ease	*www.armstrongair.com*	972.497.5000
Armstrong Pumps	Pumps	*www.armstrongpumps.com*	716.693.8813
Arsco Manufacturing	Air handler units	*www.arscomfg.com*	800.543.7040
Aurora Pump	Pumps	*www.aurorapump.com*	630.859.7000
Baltimore AirCoil	(BAC) Cooling towers	*www.baltaircoil.com*	410.799.6200
Bananza Air Mgmt.Sys.	Unit manufacturer	*www.bananza.com*	800.255.3416
Bard Manufacturing Co.	AC, heat pumps, furnaces	*www.bardhvac.com*	419.636.1194
Barry Blower	Fans/blowers	*www.barryblower.com*	763.571.5200
Bell & Gossett	(ITT) Pumps	*www.bellgossett.com*	847.966.3700

Company	Products	Web Address	Phone
Berner Corporation	Air doors & soft duct	*www.airdoors.com*	800.245.4455
Bessamaire	Gas/oil heaters	*www.bessamaire.com*	800.321.5992
Big Fans	Fans	*www.bigassfans.com*	877.BIG.FANS
Boiler Source	Boilers	*www.boilersource.com*	847.253.1040
Broad	Absorbtion chillers	*www.broad.com*	201.678.3010
Bry-Air	Moisture removal	*www.bry-air.com*	740.965.2974
Bryan Boilers	Boilers	*www.bryanboilers.com*	765.473.6651
Bryant Heating & Cooling	HVAC products	*www.bryant.com*	800.428.4326
Buderus Boilers	Boilers	*www.buderus.net*	603.552.1100
Burnham	Commercial boilers	*www.burnhamcommercial.com*	717.397.4701
Calmac	Thermal storage	*www.calmac.com*	201.569.0420
Cambridge Engineering	Special air systems	*www.cambridge-eng.com*	800.318.0661
Camus Hydronics Ltd	Hydronic boilers	*www.camus-hydronics.com*	905·696·7800
Carnes Company	Air distribution & humidifiers	*www.carnes.com*	608.845.6411
Carrier	Commercial AC equipment	*www.commercial.carrier.com*	800.227.7437
Central Boiler	Boilers	*www.centralboiler.com*	800.248.4681
Century	Unit manufacturer	*www.rae-corp.com*	918.825.7222
Refrigeration (RAECorp)	HVAC/R products mfgr.	*www.rae-corp.com*	See Web site
Cesco	Louvers/dampers/vents	*www.cescoproducts.com*	888.422.3726
Chicago Blower Corp.	Fans/blowers	*www.chiblo.com*	630.858.2600
Chromalox (Wiegand/Emerson)	boilers/controllers/sensors	*www.chromalox.com*	800.443.2640
Ciat (France)	Heat pump/equip. mfgr.	*www.ciat.fr*	(0)4.79.42.42.42
Climate Master	Water-source heat pumps	*www.climatemaster.com*	405.745.6000
Coil Company	HVAC coils	*www.coilcompany.com*	800.523.7590
Coleman AC	(York) AC units	*www.colemanac.com*	877.874.7378
Colmac Coil	Coils & fluid coolers	*www.colmaccoil.com*	509.684.2595
Comair Rotron	Fans & blowers	*www.comairrotron.com*	619.661.6688
Compu-Aire	Computer AHUs	*www.compu-aire.com*	562.945.8971
Continental Fan	Fans & blowers	*www.continentalfan.com*	800.779.4021
Convair Cooler Corp.	Evaporative coolers	*www.emersonclimate.com*	602.353.8066
Copeland Corp.	(Emerson) compressors	*www.copeland-corp.com*	317.596.9499
Cozy Heaters	Gas heaters	*www.cozyheaters.com*	502.589.5380
Crowley Fan	(Multi-Wing) fans & blower	*www.multi-wing.com*	440.834.9400

Company	Products	Web Address	Phone
Daikin, LTD	HVAC products	*www.daikin.com*	212.935.4890
Data Aire, Inc.	Computer room AC	*www.dataaire.com*	714.921.6000
Dectron International	HVAC products	*www.dectron.com*	888.332.8766
Dehumidifier Corp.	Dehumidification	*www.dehumidifiercorp.com*	888.883.7602
Delta Cooling Towers	Cooling towers	*www.deltacooling.com*	800.289.3358
Des Champs Technology	Dehumidification	*www.des-champs.com*	540.291.1111
Desert Aire	Dehumidification	*www.desert-aire.com*	414.357.7400
Detroit Radiant Products	Radiant heating	*www.reverberray.com*	586.756.0950
Donco Air Products	Air distribution	*www.doncoair.com*	641.488.2211
Dry Air (Enviro. Pool Systems, Inc.)	Pool dehumidification	*www.dry-air.com*	800.671.9629
Dri-Steem Humidifier Co.	Humidifiers	*www.dristeem.com*	800.328.4447
Ductmate Industries, Inc.	Duct hanging systems	*www.ductmate.com*	800.245.3188
Dumont Eng. Mech.Sys.	Dehumidification systems	*www.dumontgroup.com*	207.933.4811
Dunham-Bush Inc.	Chillers	*www.dunham-bush.com*	800.386.2874
Dynaforce	Air curtains	*www.dynaforce.com*	800.421.1266
EasyHeat	Radiant heating	*www.easyheat.com*	860.653.1600
Easco Boiler Corp.	Boilers	*www.easco.com*	718.378.3000
Ecosaire (Hiross)	Air handlers	*www.dectron.com*	800.387.1930
EER Products	HVAC products	*www.eerproducts.com*	888.899.9991
Energy Dynamics	Geothermal heat pump mfgr.	*www.energy-dynamics.com*	800.444.8583
Engineered Air	Custom air handlers	*www.engineeredair.com*	913.583.3181
Enviro-Tec	VAVs/FCs/AHUs	*www.enviro-tec.com*	727.541.3531
EnviroVent/Farmer PLC (UK)	Air distribution	*www.envirovent.com*	1423.81081
Ergomax	Hydronic heating	*www.ergomax.com*	908.281.1005
Evapco	Cooling towers	*www.evapco.com*	410.756.2600
EWC Controls	Zone control/ humidifiers	*www.ewccontrols.com*	800.446.3110
Fan America, Inc.	Non-corrosive fans	*www.fanam.com*	800.838.4074
Fedders Engineered Products	HVAC products	*www.fedders.com*	908.604.8686
First Company	HVAC equipment mfgr.	*www.firstco.com*	214.388.5751
Flat Plate, Inc.	Heat exchangers	*www.flatplate.com*	800.774.0474
Floor Heatech, Inc.	In-floor electric heating	*www.easywarmfloor.com*	888.894.0807
Florida Heat Pump	Heat pump manufacturer	*www.fhp-mfg.com*	See Web site

Company	Products	Web Address	Phone
Friedrich A/C Products	Room air conditioner mfgr.	www.friedrich.com	210.357.4400
Frigidaire	AC & HP units	www.frigidaire.net	800.374.4432
Fujitsu General	Split AC	www.fujitsugeneral.com	888.888.3424
Fulton	Packaged boilers	www.fulton.com	315.298.5121
Goettl	AC equipment	www.goettl.com	800.334.6494
Goodman A/C & Htg.	AC units	www.goodmanmfg.com	866.304.0137
Governair	Temtrol, Broan, Nutone	www.governair.com	405.525.6546
Greenheck Fan Corporation	Fans, blowers	www.greenheck.com	715.359.6171
GRI Pumps	Pumps	www.gripumps.com	419.886.3001
Hart & Cooley Inc.	Air distribution	www.hartandcooley.com	800.433.6341
Hartzell Fan, Inc.	Fans	www.hartzellfan.com	800.336.3267
HeatCraft (Luvata)	Refrigeration products	www.heatcraft.com	800.225.4328
Heil Products	AC equipment	www.heil-hvac.com	See Website
Herrmidifier	Humidifiers	www.herrmidifier.com	800.884.0002
HuntAir	Custom air handlers	www.huntair.com	503.639.0113
Hurst Boiler	Boilers	www.hurstboiler.com	877.99HURST
Hydro-Air Components, Inc.	Rittling/hydronics	www.hydro-air.net	800.346.8823
HydroTherm	Modular boilers	www.hydrotherm.com	413.564.5515
Indiana Fan	Fans & fan parts	www.indiana-fan.net	317.634.7165
Industrial Acoustics	Sound control products	www.industrialacoustics.com	718.931.8000
International Environmental Corp.	AC equipment	www.iec-okc.com	See Web site
Islandaire	Computer room AC	www.islandaire.com	800.886.2759
JLC International	Humidity/velocity sensors	www.jlcinternational.com	800.599.4732
Kathabar Inc.	Dehumidification	www.kathabar.com	732.356.6000
KeepRite (Canada)	Package AC units	www.keeprite.com	See Web site
Krueger	Air distribution	www.krueger-hvac.com	972.918.8269
Lakos/Laval	Separators	www.lakos-laval.com	800.344.7205
Lattner Boiler Mfg. Co.	Steam boilers	www.lattner.com	800.345.1527
Leader Fan Ind. (Canada)	Fans	www.leaderfan.com	416.675.4700
Legacy Chillers Systems	Process chillers	www.legacychillers.com	877-988-5464
Leibert (Emerson)	Computer room AC	www.leibert.com	800.543.2378
Lennox Industries Inc.	AC equipment	www.lennox.com	972.497.5000
Lindab USA	Duct systems	www.lindabusa.com	203.325.4666

Company	Products	Web Address	Phone
Lochinvar	Boilers/water heaters	www.lochinvar.com	615.889.8900
Loren Cook Company	Fans & blowers	www.lorencook.com	417.869.6474
Magic Aire (United Electric)	Coils & AHUs	www.magicaire.com	940.767.8333
Magic-Pak (Armstrong)	Package HVAC units	www.magic-pak.com	866.282.7257
Mammoth Inc.	Chillers, AC, heat pumps	www.mammoth-inc.com	800.328.3321
Markel (TPI Corp.)	Electric heat products	www.markel-products.com	800.682.3398
Marley (SPX)	Cooling towers	www.spxcooling.com	913.664.7400
Marlo Coil (DRS)	HVAC equipment	www.marlocoil.com	636.677.6600
Mars Air Products	Air systems	www.marsair.com	800.421.1266
McQuay International	Chillers, heat pumps	www.mcquay.com	612.553.5330
Mechanovent (NYB)	Fans & blowers	www.mechanovent.com	219.325.6788
Menerga Energy Systems	Dehumidification systems	www.menerga.ltd.uk	1926.62177
MetalAire	Air distribution	www.metalaire.com	727.441.2651
Metal Form Manufacturing	Dampers/louvers/ vents/sound	www.mfmca.com	602.233.1211
Mitsubishi Climate Control	DX terminal units	www.mitsubishicc.com	800.835.4494
MK Plastics Corp.	Corrosive exhausts systems	www.mkplastics.com	888.287.9988
Modine	Furnaces & make-up air units	www.modine.com	See Web site
MovinCool	Portable AC	www.movincool.com	800.264.9573
Mr. Slim	Mitsubishi HVAC	www.mrslim.com	502.239.2689
MSP Technology, LLC	Coils/HX/ humidifiers	www.msptechnology.com	877.MSP.MSP6
MTH Pumps	Pumps	www.mthpumps.com	630.552.4115
Munters	Custom air handlers	www.munters.com	800.MUNTERS
Mydax	Chillers	www.mydax.com	503.888.6662
Nailor Industries	Unit manufacturer	www.nailor.com	281.590.1172
Nederman, Inc.	Exhaust systems	www.nederman.com	800.575.0609
New York Blower Co.	Fans, blowers	www.nyb.com	630.794.5700
Nesbitt Aire	Unit ventilators/ heaters/coils	www.nesbittaire.com	216.451.9300
Norfab	Ducting systems	www.norfab.com	800.532.0830
Orival	Water filters	www.orival.com	800.567.9767
Paco Pumps	Pumps	www.paco-pumps.com	909.594.9959
Parker Boilers	Boilers	www.parkerboiler.com	323.727.9800
Parker Davis	HVAC systems pioneer/infiniti HVAC	www.pd-hvac.com	305.513.4488

Company	Products	Web Address	Phone
Peerless Heaters/ Boilers	Heaters/boilers	www.peerless-heater.com	See Web site
Penn Separator Corp.	Steam separators/ boiler assy	www.pennseparator.com	888.PENN.SEP
Penn Ventilator (PennBarry)	Fans & blowers	www.pennbarry.com	See Web site
Pentair Pump Group	Pumps	www.pentairpump.com	630.859.7000
Paharpur (India)	Cooling towers	www.paharpur.com	+91.33.24792050
PlymoVent	Exhaust systems	www.plymovent.com	732.417.0808
Poolpak International	Pool dehumidification	www.poolpak.com	800.959.PPAK
Pottorff Dampers	Dampers & louvers	www.pottorff-hvac.com	817.831.7038
Powered Aire, Inc.	Air curtains	www.poweredaire.com	888.321.AIRE
Price HVAC	Air distribution	www.price-hvac.com	770.623.8050
Pump-Flo Solutions	Pump sizing & selection	www.pump-flo.com	800.768.8545
Pure Humidifier Co.	Humidifiers	www.purehumidifier.com	952.368.9335
Radiator Corp. (Dunkirk/ECR)	Boilers	www.dunkirk.com	716.366.5500
RAE Corp.	HVAC equipment	www.rae-corp.com	918.825.7222
Raypak (Rheem)	Boilers	www.raypak.com	818.889.1500
RBI Water Heaters	Water heaters/ boilers	www.rbiwaterheaters.com	413.564.5515
Recold	Condensers/ coolers/towers	www.recold.com	714.529.6080
Redi Controls	HVAC/R maintenance equipment	www.redicontrols.com	800.626.8640
REMA Dri-Vac Corp.	Boiler returns/ tanks	www.boilerreturns.com	203.847.2464
Reznor (Thomas & Betts)	Air systems/unit heaters	www.reznoronline.com	800.695.1901
Rheem Mfg. Co. (Paloma -Japan)	AC equipment	www.rheem.com	800.548.7433
Rhoss (Italy)	AC equipment	www.rhoss.com	(+39)0432.911518
Riello Burners	Gas/oil burners	www.riello-burners.com	800.4RIELLO
Rink Sound	Sound silencers	www.rinksound.com	816.761.7476
Roberts Gordon, Inc.	Infrared heating	www.rg-inc.com	800.828.7450
Ruud (Paloma Corp.-Japan)	Ruud/rheem AC equipment	www.ruud.com	800.848.7883
Samsung DVM	DVM systems	www.quietside.com	717.243.2535
Sanyo Fisher Co.	DX terminal units	www.sanyohvac.com	800.421.5013
John C. Schaub, Inc.	Chillers	www.chillers.com	856.235.2120
Selkirk Metalbestos	Ducting/venting	www.selkirkinc.com	800.992.8368
SEMCO Inc.	Energy recovery	www.semcoinc.com	888.4.SEMCOINC

Company	Products	Web Address	Phone
Skil-Aire	Unit manufacturer	*www.skil-aire.com*	800.625.7545
Skymark International	AC equipment	*www.skymarkinternational.com*	888.SKYMARK
Slant-Fin	Hydronic heating	*www.slantfin.com*	516.484.2600
Smith Cast Iron Boilers	Cast iron boilers	*www.smithboiler.com*	413.562.9631
SPEC-AIR	Evaporative coolers	*www.specair.net*	800.288.0892
Sta-Rite Industries	Pumps	*www.starite.com*	800.472.0884
Sterling (Sterlco)	Process clg. & htg.	*www.sterlco.com*	414.354.0970
Sterling HVAC (Mestek)	Hydronic heating	*www.sterlinghvac.com*	413.564.5540
Stirling (UltimateAir)	Residential energy recovery	*www.ultimateair.com*	800.535.3448
Strobic Air	Fans & lab. exhaust	*www.strobicair.com*	800.722.3267
Stulz ATS	Humidifiers/ dehumidifiers	*http://ats.stulz.com*	301.663.8885
Superior Boiler Works (Canada)	Boilers	*www.superiorboilerworks.com*	905.561.1413
Superior Rex Fancoils	Fancoil manufacturer	*www.superiorrex.com*	954.457.7778
Super Radiator Coils	Coils	*www.srcoils.com*	800.394.2645
Swartwout	Fiberglass/PVC dampers	*www.swartwout.com*	816.761.7476
TACO HVAC	Pumps, heat exchangers	*www.taco-hvac.com*	815.344.3766
Tandem Chillers	Modular chillers	*www.tandomchillers.com*	877.513.8330
TC Vent Co.	Fans	*www.tcventco.com*	763.551.7600
Tecogen	Gas engine driven chillers	*www.tecogen.com*	781.622.1400
Temtrol	Coils, AHUs	*www.temtrol.com*	405.263.7286
Thermo Pride	AC & furnaces	*www.thermopride.com*	574.896.2133
Thermal Solutions	Boilers	*www.thermalsolutions.com*	717.293.5811
Titus	Air distribution	*www.titus-hvac.com*	972.699.1030
Tjernlund Products, Inc.	HVAC products	*www.tjernlund.com*	800.255.4208
Tower Tech	Cooling towers	*www.towertechinc.com*	405.290.7788
Tranter, Inc. (UK)	Heat exchangers	*www.tranter.com*	940.723.7125
Triad Boiler	Boilers	*www.triadboiler.com*	630.562.2700
Tri-Mer Corp.	Scrubbers	*www.tri-mer.com*	517.723.7838
Tristate Coil	Coils	*www.tristatecoil.com*	See Website
Trion Inc.	Air purification	*www.trioninc.com*	800.884.0002
Troy Boiler Works	Boilers	*www.troyboilerworks.com*	518.274.2650
Tuscan Foundry (UK)	Cast iron radiators	*www.tuscan-build.com*	1403.865028
Tuttle & Bailey	Air distribution	*www.tuttleandbailey.com*	972.680.9128
Twin City Fan	Fans	*www.twincityfan.com*	763.551.7600
Unilux Boiler Corp.	Boilers	*www.uniluxboilers.com*	905.851.3981
United Cool Air Corp.	Unit manufacturer	*www.unitedcoolair.com*	877.905.1111

Company	Products	Web Address	Phone
United Metal Products	Unit manufacturer	*www.unitedmetal.com*	480.968.9550
USA Coil	Coils	*www.usacoil.com*	800.872.2645
Venmar (Canada)	Ventilation	*www.venmar.ca*	800.567.3855
Vent-A-Fume	Fume venting systems	*www.ventafume.com*	716.876.2023
Ventex Inc. (Canada)	Dampers & louvers	*www.ventexinc.com*	800.668.7214
Ventrol (CES Group) Custom	AHUs	*www.ventrol.com*	514.354.7776
Vikrant International (India)	Compressors & components	*www.vikrantindia.com*	+91.22.62372051/ 6243729
Water Furnace	Geothermal heat pumps	*www.waterfurnace.com*	260.479.3225
Weil-McLain	Hydronic heating	*www.weil-mclain.com*	219.879.6561
YORK (Johnson Controls)	Chillers, AHUs	*www.york.com*	717.771.7890

Commercial Refrigerator Manufacturers

Company	Brand Name
Beverage-Air Corp.	Beverage-Air
Blue Air Commercial Refrigeration, Inc.	Blue Air
BuSung America Corp.	Everest Refrigeration
Continental Refrigerator	Continental
Duke Manufacturing	Backcounter
Electrolux Home Products	Artic-Air, Frigidaire, Kelvinator
Everest Import Distribution	Global 1st Products
Fogel de Centroamerica S.A.	Fogel
Frigoglass North America	Miracool by Frigoglass
Hoshizaki America, Inc.	Hoshizaki
Imbera S.A. de C.V.	Imbera
Innovative Displayworks, Inc.	IDW
Jiangsu Baixue Electric Appliances Co., Ltd.	AHT, Baixue
Jimex Corp.	Ascend, Jimex Ascend
Master-Bilt Products	Master-Bilt
McCall Refrigeration	McCall
Metalfrio Solutions	Metalfrio Solutions, Inc.
Nor-Lake Incorporated	Norlake
Patriot Food Service Equipment	Patriot Food Service Equipment
Perlick Corporation	Perlick
Primoris Inc., Saturn Equipment	Primoris Inc.
QBD Cooling Systems Inc.	QBD
Qingdao Haier Special Icebox Co., Ltd	Haier
Royal Vendors Inc.	Royal Vendors
SG Beverage Solutions	SG Beverage

Company	Brand Name
The Delfield Company	Delfield
Traulsen	Traulsen
True Manufacturing (True Food Service Equipment)	True Refrigeration
Turbo Air, Inc.	Turbo Air
Victory Refrigeration	Victory

Commercial Freezer Manufacturers

Company	Brand Name
Beverage-Air Corp.	Beverage-Air
Blue Air Commercial Refrigeration Inc.	Blue Air
Continental Refrigerator	Continental
Electrolux Home Products	Artic Air, Frigidaire, Kelvinator
Hoshizaki America, Inc.	Hoshizaki
Jimex Corp.	Ascend, Ascend by Jimex Corp
Master-Bilt	Master-Bilt
McCall Refrigeration	McCall
Metalfrio Solutions	Metalfrio Solutions, Inc.
Nor-Lake Incorporated	Norlake
Patroit Food Service Equipment	Patroit Food Service Equipment
Primoris Inc. Saturn Equipment	Primoris Inc.
The Delfield Company	Delfield
Traulsen	Traulsen
True Manufacturing (True Food Service Equipment)	True Refrigeration
Turbo Air, Inc.	Turbo Air
Victory Refrigeration	Victory

Commercial Ice Machine Manufacturer's

Company	Web Address
A&V Refrigeration	*www.av-refrigeration.com*
A1 Refrigeration	*www.A1flakeice.com*
Cornelius	*www.cornelius-usa.com*
Follett	*www.follettice.com*
GEA Refrigeration	*www.gearefrigeration.com*
Holiday Ice	*www.holiday-ice.com*
Hoshizaki	*www.hoshizakiamerica.com*
Ice-O-Matic	*www.iceomatic.com*
Kold-Draft	*www.kold-draft.com*
Manitowac	*www.manitowocice.com*
North Star	*www.northstarrice.com*
Scotsman	*www.scotsmanice.com*
Snowell Ice Systems	*www.flake-icemachine.com*
U.S. Ice Machine Mfg. Co.	*www.usicemachine.com*
Vogt	*www.vogtice.com*

Supermarket Refrigeration Manufacturers

Company	Web Address
Hill Phoenix	*www.hillphoenix.com*
Hussman	*www.hussman.com*
Kysor-Warren	*www.kysorwarren.com*
Magma	*www.magmacold.com*
Piper Products (Servolift/Eastern)	*www.piperonline.net*
Utility Refrigerator	*www.utilityrefrigerator.com*
Zero-Zero Refrigerator Mfg. Co.	*www.zero-zero.com*

Glossary

aldehyde   A class of compounds that can be produced from incomplete combustion of gas fuels.

allen wrench   An L-shaped tool that is used to remove hex screws.

alternating current (AC)   Electrical current that flows in one direction and then reverses itself and flows in the opposite direction. In 60-cycle current, the direction of flow reverses every 120th second.

ambient temperature   Temperature of air that surrounds an object on all sides.

ammeter   A test instrument used to measure current.

ampere   The number of electrons passing a given point in 1 second.

belt   A band of flexible material used to transfer mechanical power from one pulley to another.

blower wheel   A device attached to the indoor side of the fan motor shaft and is used to circulate air across the evaporator coil.

bracket   A hardened structure used to support a component.

BTU   (British thermal unit) The amount of heat required to raise the temperature of 1 lb of water 1°F.

burner   A device in which an air and gas mixture has entered into the combustion zone to be expelled and burned off.

cabinet   The outer wrapper of an appliance.

capacitor   A device that stores electricity; used to start and/or run circuits on many large electric motors. The capacitor reduces line current and steadies the voltage supply, while greatly improving the torque characteristics of the fan motor or compressor.

capillary tube   A metering device used to control the flow of refrigerant in a sealed system. It usually consists of several feet of tubing with a small inside diameter.

celsius   The metric system for measuring temperature. The interval between the freezing point and the boiling point of water is divided into 100°. This is also called the centigrade scale.

circuit   A path for electrical current to flow from the power source to the point of use and back to the power source.

circuit breaker   A safety device used to open a circuit if that circuit is overloaded.

closed (circuit)   An electrical circuit in which electrons are flowing.

combustion   The rapid chemical reaction of a gas with oxygen to produce heat and light.

component   An individual mechanical or electrical part of an appliance.

compressor   An apparatus similar to a pump that is used to increase the pressure of refrigerant that is to be circulated within a closed system.

condenser coil   A component of the refrigeration system that transfers the heat that comes from within a closed compartment to the outside surrounding area.

condenser fan   A fan motor and fan blades used to cool the condenser coil.

contact points   Two movable objects, or contacts, that come together to complete a circuit or that separate and break a circuit. These contact points are usually made of silver.

continuity   The ability of a completed circuit to conduct electricity.

control   A device, either automatic or manual, that is used to start, stop, or regulate the flow of liquid, gas, or electricity.

current   The flow of electrons from a negative to a positive potential.

cycle   A series of events that repeat themselves in the same order.

defective   Refers to a component that does not function properly.

defrost cycle   That part of the refrigeration cycle in which the ice is melted off the evaporator coil.

defrost thermostat (heat pump AC models)   A dual-purpose control that acts as an outdoor thermostat and defrost control.

defrost timer   A device used in an electrical circuit to turn off the refrigeration long enough to permit the ice to be melted off the evaporator coil.

diagnosis   The act of identifying a problem based on its signs and symptoms.

direct current (DC)   Electric current that flows in one direction in a circuit.

energize   To supply electrical current for operation of an electrical component.

evaporator coil   A component of the refrigeration system that removes the heat from within a closed compartment.

evaporator fan   A motor and fan blades used to circulate the cold air.

fahrenheit   The standard system for measuring temperature. The freezing point of water is 32°F, and the boiling point of water is 212°F.

flashback   Also known as extinction pop. Where the burner flames enter back into the burner head and continue to burn after the gas supply has been turned off.

fuse   A safety device used to open a circuit if that circuit is overloaded.

gasket   A flexible material (either airtight or watertight) used to seal components together.

ground   A connection to earth or to another conducting body that transmits current to earth.

ground wire   An electrical wire that will safely conduct electricity from a structure to earth.

heat anticipator   Used to provide better thermostat and room air temperature control.

hertz (Hz)   A unit of measurement for frequency. One hertz equals one cycle per second.

hot gas defrost   A defrosting system in which hot refrigerant from the condenser coil is directed to the evaporator coil for a short period of time to melt the ice from the evaporator coil.

housing   A metal or plastic casing that covers a component.

insulation   Substance used to retard or slow down the flow of heat through a substance.

insulator   A material that does not conduct electricity. It is used to isolate current-carrying wires or components from other metal parts.

ladder diagram   A wiring schematic in which all of the components are stacked in the form of a ladder.

lead (wire)   A section of electrical wiring that is attached to a component.

module   A self-contained device with a group of interconnecting parts designed to do a specific job.

nut driver   A tool used to remove or reinstall hexagonal-head screws or nuts.

ohm   A unit of measurement of electrical resistance.

ohmmeter   A test instrument used for measuring resistance.

open (circuit)   A break in an electrical circuit that stops the flow of current.

orifice   An opening where the gas is metered.

overload protector   A device that is temperature- , pressure- , or current-operated that is used to open a circuit to stop the operation of that circuit should dangerous conditions arise.

parallel circuit   Components that are parallel-connected across one voltage source. All of the branches are supplied with the same amount of voltage.

pressure switch   A device that is operated by pressure and turns a component on or off.

primary air   The main air supply introduced into the burner which mixes with the gas before it reaches the burner head.

PTC (positive temperature coefficient)   A resistor that increases resistance (ohms) with temperature increase.

pulley   A wheel that is turned by or driven by a belt.

refrigerant   A chemical substance used in refrigeration and air conditioning that produces a cooling effect.

relay   A magnetic switch that uses a small amount of current in the control circuit to operate a component needing a larger amount of current in the operating circuit. A remote switch.

relief valve   A safety device that is designed to open before dangerous pressure is reached.

resistance   The opposition to current flow. The load in an electrical circuit.

run winding   Electrical winding of a motor, which has current flowing through it during normal operation of the motor.

safety thermostat   A thermostat that limits the temperature to a safe level.

schematic diagram   A line drawing that gives the electrical paths, layout of components, terminal identification, color codes of wiring, and sometimes the sequence of events of an appliance.

series circuit   A circuit in which all of the components are sequentially connected in the same line. If one component fails, all the components fail.

short circuit   An electrical condition in which part of a circuit errantly touches another part of a circuit and causes part or all of the circuit to fail (and trip the circuit breaker or blow a fuse).

slinger ring   A ring attached to the circumference of the condenser fan blade. The slinger ring is positioned close to the base pan of the air conditioner so water in the base pan is picked up by the rotating ring and dispersed into the hot condenser air stream where it evaporates.

solenoid   A cylindrical coil of insulated wire that establishes a magnetic field in the presence of current.

start winding   A winding in an electric motor used only during brief periods when the motor is starting.

switch   A device to turn current on or off in an electrical circuit.

temperature   A measure of heat energy or the relative lack thereof.

terminal   A connecting point in a circuit to which a wire would be attached to connect a component.

test light   A light provided with test leads that is used to test electrical circuits.

thermistor   A device that exhibits a large change in electrical resistance with a change in temperature.

thermometer   A device used to measure temperature.

thermostat   A device that senses temperature changes and that usually operates a control relay.

thermostat (operating)   A thermostat that controls the operating temperature of a component.

transformer   A device that raises or lowers the main AC supply voltage.

voltage   The difference in potential between two points; the difference in static charges between two points.

voltmeter   A test instrument used to measure voltage.

VOM (volt-ohm-multimeter)   A test instrument used to measure voltage, resistance, and amperage.

water column (WC)   A unit used for pressure. A 1-in water column equals a pressure of 0.578 oz/ in^2.

watt   A unit to measure electrical power.

wattmeter   A test instrument used to measure electrical power.

Index

Note: Page numbers followed by *f* denote figures; page numbers followed by *t* denote tables.